ENVIRONMENTAL JUSTICE in NEW MEXICO

Counting Coup

VALERIE RANGEL

Foreword by Bob Haozous

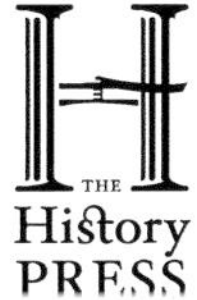

Published by The History Press
Charleston, SC
www.historypress.com

Front cover, clockwise from upper left: Oil painting *Home Sweet Home. Craig Nez George, 2016*; Cleanup of a nuclear uranium waste spill along the Rio Puerco, on the Navajo Nation, 1979. *Photograph: Dan Budnik. © [1979] Dan Budnik. All rights reserved*; Horse that died from drinking water contaminated by the nuclear spill into the Rio Puerco, 1979. *Photograph: Dan Budnik. © [1979] Dan Budnik. All rights reserved*; Protect Mount Taylor mural. *Photo and art by Nani Chacon*; Tularosa Basin Downwinders Consortium demonstration. *Photo: Tina Cordova*; First atomic bomb detonated on July 16, 1945, at the Trinity site, New Mexico. *Photo*: *Center for Southwest Research, University Libraries, University of New Mexico.*
Back cover: Trash in the Rio Grande. *Photo: AMAFCA*; *insert*: Papercut image of Pesticide Drift. *Image by Valerie Rangel.*

First published 2019
Updated printing 2024

Manufactured in the United States

ISBN 9781467141338

Library of Congress Control Number: 2018960966

CONTENTS

CONTENTS

FOREWORD

Modern Indigenous life has become immersed with Western society's contemporary perspectives. Language, religion, politics, law, the arts or any of past Indigenous social systems are now held firm captive by this Western point of view. The identity assumptions of native peoples long sanctioned as definitive have become dated, inadequate and lacking of purpose. As contemporary people, we must now seek an updated and honest redefinition. Function and purpose have always been the vital and essential basis to this reevaluated identity. Who we were or what we once believed must not dominate any new contemporary self-description. If this renewed identity is to be meaningful, it must have Indigenous nature relationships as its basis.

The picking, choosing and merging with pan-Indian/Indigenous concepts and other popular Indianisms have gradually become an expedient and feel-good form of cultural retention.

Purchasers of Indian art products are knowingly restricting intellectual input by supporting an art market whose only value is the buyer's taste and the economic value of the product. This influential demand for pretty, romantic and Indian products dramatically reduces the potential of any valuable internal dialog. Western-trained academics unwittingly support this cosmetic Indian identity. Economic reasoning and superficial interpretations support the demand for feathers, smoke, prayer, costume, decoration, traditions and history. Monetary necessities are forcing contemporary Indigenous artists to dismiss environmental responsibilities to satisfy a market unsupportive of honest introspection. The short-term economic rewards of art for vocation leave our distant descendants with a questionable inheritance. This desire to gain the modern world's conveniences implies our support and blessing for this disassociation from nature. The traditional call to action for Indigenous man was the "stewards of nature" role. This inspirational-sounding Indigenous cliché has now become generically Indian. Understanding the methods of colonization used to separate Indigenous peoples from their cultural roots is important. First was the removal of children from their families, traditional land and cultural base. These captive children were given an extensive reeducation based on Western language, science, mathematics, medicine and law. This teaching intentionally replaced the pagan-like spirituality and of-the-earth wisdom with an economic-based belief system. We were taught that contemporary problems were beyond the capability of Indigenous solutions and that these Western rationalities were the only effective means of adapting to the modern world. Children and adults were effectively taught that the personal gain from individualism was far more important than tribal efforts or allegiances. Faint embers of what little tribal memory remaining were very close to being extinguished. The children were told to dismiss previous tribalism, culture and wisdom as "of the past and extinct." Our goal was to convince the "white" man that we could adapt and succeed in all efforts of mimicry by mastering their medicine, law, science, military service and religion. The success of these goals was devastating to traditional native life ways. This form of trickle-down prestige offered nothing of substance to share with one's people, place or tribe. Western colonization has always been reliant upon economic goals for its rationale.

Economic logic has overruled the primary Indigenous purpose once dependent on nature and its laws. This now-ingrained education has become so integral to Indigenous identity that questioning it renounces the speaker as iconoclastic and contrary. Our participation in Western society helps maintain the continuation of colonization while simultaneously dismissing the primary legacy of environmental caretaking.

Our own indoctrinated educators are most often teachers of these contemporary political dogmas.

This method of support is maintained with patriotism, loyalty, faith, belief and the trust in Western man's procedures. In the contemporary world, a massive fracture now exists between economic and environmental responsibilities.

Capitulation to the Western religious beliefs that remove us from this "hell on earth" at the end of our lives demands mimicry of Western teaching by proclaiming that mother earth's sole benefactor is mankind. The immediate effectiveness of "new world" American pragmatism has created dramatic short-term successes that leave severe and continuing long-term problems for the planet.

In the dreamer's world, the creation of a perfect world from an Indigenous framework would be the maintenance of the balance of all things. This could be a goal for all future life on this planet if we take our purpose seriously. In today's political and environmental climate, change is awaiting Indigenous leadership.

To an Indigenous way of thinking, dissention was a chance for contemplation, adaptation and compromise with the aim of meaningful and mutually beneficial solutions. The traditional method of confrontation or "counting coup" (the use of humor and ridicule to confront, diffuse and empower all issues within communities) has been altered into Western goals of winning, defeating, controlling or destroying. A fundamental component of our identity is stated in the concept of "all our relations." This common understanding referenced the two-legged and the four-legged, the things that fly and crawl, the things that live in the water and in the ground, not excluding stones, rain, sunlight and literally everything of the universe mentioned in this holistic concept.

Counting coup on American economic democracy with the desire for healthy solutions and change can only benefit all things.

BOB HAOZOUS
http://www.bobhaozous.com

Bob Haozous was born in Santa Fe, New Mexico, in 1943 to Allan Houser (Chiricahua Apache) and Anna Marie Gallegos (Navajo/English/Spanish). He grew up in northern Utah, where his parents were teachers at the Inter Mountain Indian School in Brigham City. Haozous studied at Utah State University before enlisting in the U.S. Navy, where he served for four years on board the USS Frank Knox *during the Vietnam War. After the war, Haozous attended the California College of Arts and Crafts in Oakland, California, where he earned his BFA degree in sculpture in 1971. Haozous is one of the most important Native sculptors of the Native American Fine Art Movement. His innovation and experimentations with materials push the boundaries of "Indian" art—the boundaries that his father helped to define. He is best known for his monumental cut steel pieces, which often deal with poignant topical issues. He approaches these issues with a bit of a bite and a good dose of humor. His injection of humor allows the serious issues to be more palatable and to have a universal presence.*

FOREWORD

Haozous has chosen to take back his Apache family name and to reject the Anglo version: Houser. This name was given to his father as a child in an Oklahoma Indian boarding school. Together, Haozous and Houser represent the breadth and depth of Native American sculpture. Haozous has been able to establish himself as a leading artist because of his father's encouragement and nurturing. As well, Haozous has encouraged and supported his father's work.

ACKNOWLEDGEMENTS

This book is dedicated to my mother, whose life, struggles and passing inspired this book.

Special appreciation, gratitude and respect to Winona LaDuke and the Honor the Earth organization.

Thank you to Earl Tully, Lori Goodman, Daniel Tso, Evalyn Bemis, Craig George, Nani Chacon, Frost Fowler, Jerry Lovato (AMAFCA), Daniel Lombardi, Tina Garnanez, Olivia Romo, Tina Cordova, Dr. Theodore Jojola, Dr. David Henkel, Dr. Estevan Rael-Galvez, Mary Lang, John Zambrano, Nora Naranjo Morse and Tanya Campos.

Also, a heartfelt thanks and recognition to Bob Haozous, Dan Budnick, the New Mexico Health Equity Partnership, community planners, activists and eco-organizations that so generously supported the publishing of this book of collective community voices.

INTRODUCTION

In a world full of profound and sometimes cruel ironies, one stands out: Native Americans who held their lands of the Western Hemisphere in a living trust for thousands of years, have been afflicted by some of the worst pollution of an environmental crisis that has reached planet-wide proportions. Statements from indigenous peoples around the world indicate that they perceive themselves as having been "pushed to the edge of a cliff" by environmental problems caused by industrialism.

—Melvin R.G., Prairie Smoke

For thousands of years, New Mexico was a trade hub between present-day United States, Canada and central Mexico. Political views, knowledge and stories were communicated via the Camino Real, later the Santa Fe Trail. The railroad network crisscrossed lines of communication even farther through the transportation of exotic goods into the region and out through exportation of New Mexico's natural resources. In the digital age of the new millennium, goods are bought and sold via the Internet then shipped from huge distribution warehouses directly to individual homes. Satellite imagery is an "all-seeing eye" that can spy on any aspect of the planet through visible and nonvisible spectrums.

The massive global web of knowledge and interconnectedness has never been so disconnected to the natural world as it is today. Communication is instantaneous today if you have a cell phone or can pick up Wi-Fi from a computer; however, the very devices that have brought knowledge and communication, trade goods and commerce to our fingertips have also pillaged and plummeted our environment into a sad state of disrepair. Children of urban areas grow up in a concrete jungle, disengaged from nature, exposed incrementally to procured lawns, pesticide-sprayed buildings, processed food and strategic nature outings at a city park or zoo.

To appreciate the impact of environmental racism and the magnitude of devastation in New Mexico, it is necessary to understand the land use and history, as well as the perspectives of indigenous populations and communities of color. Land, to the indigenous peoples of this continent, is not a commodity or inert material; it is "alive" and imbued with spirit! Holistic concepts of interconnectedness form the cultural worldview of Native Americans, and this land ethic continues to be passed on to subsequent generations through symbols, art, song, dance and story imbedded in language and religion. Every place is associated with spirits and legends that are remembered, revered and preserved through language and rituals.

Sometime in our recent history, the decision was made to compromise human lives and the environment; to make ecological sacrifices for a more industrialized, capitalistic world. Population growth and new technological advancements shifted the dominant society to one that is powered by an unquenched thirst for fuel, food and material possessions. Suddenly, land use planning and resource management were at the forefront of decision making and key to political control.

Planet earth is a dynamic organism, its internal processes of convective magma circulating within the planet propel land movement and atmospheric cycles, and an invisible electromagnetic outer shield protects against bombardment from incoming asteroids; without these mechanisms, our planet would cease to be "alive" and functioning. There is no place like earth in our solar system or, from what we can see, beyond. Yet, there is no place on earth that has not been impacted by man. Pristine ecosystems no longer exist, and true environmental restoration is now an impossible feat.

According to the World Health Organization (WHO), the term *ecosystem* refers to the combined physical and biological components of an environment. Organisms form complex sets of relationships and function as a unit as they interact with their physical environment. Functioning ecosystems provide invaluable services for the well-being of all organisms on the planet. These services include filtering air and water, regulating the pH of natural waters, storing fresh water, replenishing food and timber, regulating vectors, pests and pathogens, and maintaining the population balance of species.[1]

The health and happiness of mankind depends on ecological services; their functionality, integrity and sustainability stabilize the natural environment. According to the WHO, "Indirectly, changes in ecosystem services affect livelihoods, income, local migration and, on occasion, may even cause political conflict. The resultant impacts on economic and physical security, freedom, choice and social relations have wide-ranging impacts on well-being and health, and the availability and access to health services and medicines."[2]

In 2000, the Millennium Ecosystem Assessment (MA) was called for by the UN secretary general to assess the consequences of ecosystem change for human health and well-being. The MA findings show that "human actions are depleting Earth's natural capital, putting strain on the planet's ecosystems."[3] The report suggests that with appropriate actions, it is possible to reverse environmental degradation and ecosystem functions over the next fifty years with system changes and drastic policy reform however, current practices and trends show that substantial changes are not currently underway.

Due to an exponentially increasing human population, which has already caused destruction of ecosystems and significant loss of biodiversity around the globe, there is a need for "holistic planning" and innovative ways to manage resources. Apologies, lies and excuses are no longer acceptable. A renewed appreciation for the value of ecosystems and their functions is desperately needed, otherwise, environmental degradation will inevitably lead to economic instability, which can then result in civil war or unrest. Mankind must renew its commitment to conservation and remediation in order to resolve social conflicts related to land use.

I am not the creator of stories. I am the collector of painful truths and harsh realities. The justice that we envision *is possible* through the *unification* of people, *amalgamation* of knowledge and through *consilience*.[4] This book takes the reader on a journey that starts with the arrival of early colonizers, who instigated an era of landscape change that paved the way for mass resource mining and land use practices in New Mexico. The first section of this book focuses on the environmental legacy of uranium and the impact on communities that have been affected by its extraction, dumping, testing, access to water and storage. The next section focuses on environmental injustice related to water issues, highlighting the stories of the Pueblo of Isleta, the Gila River, Zuni Salt Lake, the Rio Grande, Las Animas River and the Santa Fe River. Land contamination issues such as illegal dumping, Superfund sites, immigrant farmworkers and the preservation of historic cultural resources and archaeological sites such as Chaco Canyon illustrate other types of social and environmental justice issues within the state.

PART I

LEGACY OF URANIUM

1
SCORCHED-EARTH POLICY AND MILITARISM

Early History

Evidence of the earliest prehistoric residents of New Mexico dates as far back as 25,000 BC, according to Dr. Frank C. Hibben's 1936 discovery in the Sandia Mountains. Artifacts from the Clovis Man are dated as far back as 12,000 years ago and show a transition from hunters of mammoths and bison to a plant-gatherer nomadic lifestyle after the last major ice age. Other prehistoric sites in New Mexico are located in Folsum in northern New Mexico, along the Rio Grande and in Burnett Cave west of Carlsbad.[5]

Cultivated plants such as corn, beans and squash, which originated in Mexico, are evidence that the Anasazi, Mogollon and Hohokam peoples had an extensive ecological trade network of shared goods and knowledge. Present-day tribes of Arizona believe their ancestors were the Hohokam, while the ruins of Mesa Verde, Chaco Canyon, Aztec, Bandelier and Frijoles Canyon are thought to be the ancestors of the current Pueblo tribes.[6]

By AD 400, most of the Ancient Peoples of New Mexico in what is now western New Mexico had begun to settle into villages located along cultivated river drainages. In addition to hunting and gathering of wild foodstuffs to supplement their diets, these peoples also began to develop distinctive styles of baskets and pottery. After AD 500, settlements within the western two-thirds of the state became more densely populated. Housing became more complex, with construction of towns consisting of hundreds of rooms along with specialized ceremonial structures known as kivas. Regional differences in architecture and ceramics became more pronounced as reliance on agriculture intensified and elaborate trade networks developed throughout the Southwest.

Between 1100 and 1300, many apparently prosperous and elaborate pueblos, such as Chaco Canyon, were deserted. The peoples who abandoned the ancient ruins of Chaco Canyon, Bandelier and Mesa Verde relocated along the Rio Grande and its tributaries. Colonizing people would change the environment in drastic ways, introducing plants and animals to the Southwest region, altering diets of indigenous peoples and transforming the area into an agrarian society with city centers built by the forced labor of indigenous slaves. According to historian Joe Sando, only the Hopi persisted in their ancestral Anasazi homeland, where they continue to farm a region that has no permanent rivers. Upon arrival in New Mexico, Spanish conquistadors encountered pueblos situated, as they are today, along main sources of water.[7]

History for indigenous peoples of New Mexico is bittersweet. While conquest by Spanish colonialists brought exotic goods and new technology, it also came with policies of extermination of Indian culture; suppression of language, religion and government; and slavery. Colonizers changed the environment in drastic ways; new settlers introduced foreign plants and animals to the Southwest region, altering the diet of indigenous people and transforming the landscape. City centers and churches were built with forced labor of indigenous slaves. As territorial New Mexico fell from Mexico into the hands of Anglos, "land grabbing" became a pioneering effort as was the assimilation of Indians into "Mainstream American life."[8]

Nora Naranjo Morse's public art piece *Numbe Whageh* stands in front of the Albuquerque Museum of Art and History as a reminder of the historical Pueblo occupation of those grounds prior to colonization and the battle that was fought by this Santa Clara artist to create a piece about reverence for the earth and to pay homage to the profound connection Pueblo Nations have to the land.[9] Many battles occurred in New Mexico as indigenous people and communities of color struggled to defend sacred places, their people and lands from contamination—a struggle that continues today. These stories of strife and struggle have shaped the character of the people, their traditions and their reactions to environmental and social issues in their communities.

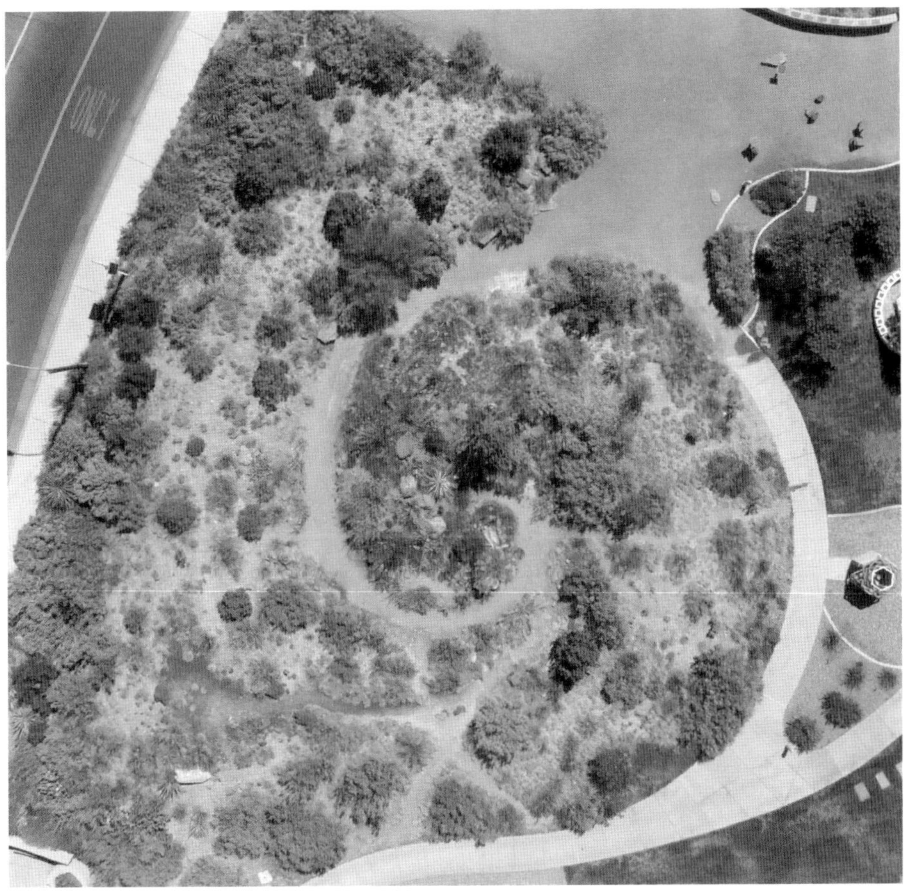

"NumbeWhageh," Albuquerque, New Mexico. *Photo credit and public art by: Nora Naranjo Morse.*

In a letter from activist, AIM leader and political prisoner Leonard Peltier dated May 13, 1976, Leonard states: "In the late 19th century, land was stolen for economic reasons. We were left with what was believed to be worthless land. Still we managed to live and defy the wish to exterminate us. Today, what was once called worthless land suddenly becomes valuable as the technology of the white society advances....[That society] would now like to push us off our reservations because beneath the barren land lie valuable mineral resources."[10]

The term *Manifest Destiny* refers to the period of time beginning in the 1800s when western expansion became the doctrine of the dominant culture in the United States. European immigrants enticed by the idea of agricultural and mining opportunities spread across the continent. Native Americans would be the focus of eradication attempts in order to ensure safe commerce and passage of settlers expanding toward the West. "Through the confiscation of Native American lands and resources as well as through alteration of natural environments, the U.S. industrial and agricultural economy grew rapidly."[11]

After the Treaty of Guadalupe Hidalgo, "land grabbing" from sparsely populated Indian settlements and borderlands of Spanish land grants by Anglo-Americans and Spanish colonists began. "By 1854, the District Court in Santa Fe ruled that under the laws of Congress, there was no Indian country in New Mexico. With one swift ruling, all Indian land was opened to New Mexican ranchers and farmers for the taking. The Native Americans viewed the new ranchers and farmers as trespassers on their land. The settlers saw the American Indian as a growing threat to their new way of life."[12]

The effects of the Civil War were especially difficult on the residents of New Mexico. According to muster rolls held at the state archives in Santa Fe, hundreds of men across New Mexico enlisted in the army, with many officers resigning or deserting due to the need to defend their family or home, to deal with economic strife and to harvest or plant crops.[13] Increasing complaints to the U.S. government regarding Indian raids on Spanish and Pueblo settlements prompted the U.S. government to turn its efforts following the war to the issue of subduing native populations and settling "untamed lands."

In 1862, during the Civil War, President Abraham Lincoln decreed freedom for all slaves through his Emancipation Proclamation and in that same year approved the establishment of Fort Sumner, which was justified by General James Carleton as offering protection to settlers in the Pecos River Valley from Mescalero Apache, Kiowa and Comanche Indians. The fort on the Bosque Redondo of the Pecos River would later become an internment camp for Navajo and the Mescalero Apache tribes.

General Carleton focused his attention on solving the "Navajo problem." Navajos were blamed for much of the raiding of food and livestock in New Mexico, and again Carleton enlisted the help of Colonel Kit Carson. On June 15, 1863, Carleton issued the order for Carson to attack the Navajo. The Navajos were in such large numbers and their country so vast that it was physically challenging to capture all. Carson knew he would fail unless he had experienced guides and trackers who knew Navajo ways and hiding places. During the winter of 1863–64, Carson employed New Mexico Volunteer Cavalry Regiments; one hundred Ute Indians, Pueblos and Hopi Indian scouts; and other informants to ravage the Navajo countryside. Carson's

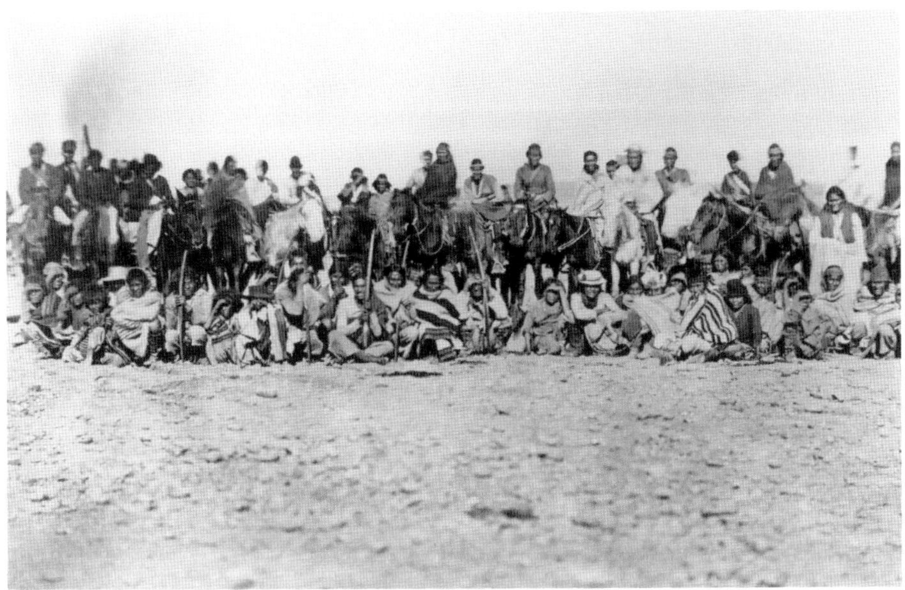

Navajo Long Walk, 1863–1867; Bosque Redondo Indian Reservation, New Mexico. *Photo credit: Center for Southwest Research, University Libraries, University of New Mexico.*

orders were to kill any Navajo resisting surrender, burn crops and orchards, kill livestock, destroy homes and contaminate water sources. This "scorched earth" policy "effectively starved the Navajo into submission."[14]

Between the summer of 1863 and the winter of 1866, 11,500 Navajo Navajos were led, by several marches, through major cities on a forced march to a destination four hundred miles away. Only an estimated 8,500 reached Fort Sumner; many Navajos who could not swim lost their lives traversing the Rio Grande. The sick and pregnant were either left by the roadside or shot and left unburied, while some women and children were captured by slave traders. Smallpox and other infectious diseases killed many incarcerated, as did exposure from inadequate clothing, lack of shelter and little to no fuel for fire.[15]

The purpose of the Bosque Redondo reservation was to force nomadic hunter-gatherer tribes of the Navajo and Apache to adopt an agrarian lifestyle and sedentary patterns similar to Pueblo Indians. By 1863, the fields that had been plowed and the crops planted failed and were not harvested due to cutworms and alkaline water. Without the intended crops, army rations were meager, resulting in the decline of health in the native populations and human suffering. Without their traditional diets, people became sick with dysentery from drinking alkaline water from the Pecos River. Army rations of rancid bacon, uncooked flour and coffee beans contributed to further stomach ailments for the indigenous population. Navajos would refer to this time and place as *Hwéeldi* ("pushed aside").[16]

In what was deemed a failed attempt at resolving the "Indian problem" in New Mexico, General Carleton was removed from command and ordered to report to duty with his regiment in Texas on February 25, 1867. In the spring of 1868, General William T. Sherman and Colonel Samuel F. Tappan met with Navajo leaders Barboncito and Manuelito to negotiate the Treaty of 1868, which established, under federal law, the sovereignty of the Navajo Nation. On June 15, 1868, the Navajos began their journey home with a cavalry escort that lasted over a month.[17] Fort Wingate was the last stop for the Navajos on the long road back to their newly established reservation. Under the treaty, parcels of land were granted to individuals and two livestock animals promised to every man, woman and child.

Bosque Redondo and the policies surrounding "Indian Removal" exemplify environmental racism and give the context for land use patterns and practices that displaced indigenous peoples from their homelands and sacred places. Relocation, eradication and forced assimilation created significant health crises such as starvation, infectious diseases, inadequate housing, mistreatment, and substantial loss of life at the hands of the military. These experiences are remembered and resonate within tribes; they are the source of deep intergenerational trauma and mistrust of the federal government today.

FORT WINGATE MILITARISM AND ENVIRONMENTAL CONTAMINATION

On October 22, 1862, General James H. Carleton ordered the establishment of the first Fort Wingate, named after Capitan Benjamin Wingate, located seventeen miles east of the present-day city of Gallup, New Mexico. In 1868, the army reactivated the Fort Lyon post at Bear Springs in order to maintain control over the Navajos returning from Fort Sumner.[18] By 1871, fifteen Navajos volunteered to fight Apaches during the last of the Indian Wars and period of Apache resistance.

From 1907 to 1912, Navajo households and two scouts applied for land allotments offered through the Dawes Act of 1887. Most of the Navajo who applied were granted land allotments within four miles west and northwest of the Fort Wingate area. In 1915, the U.S. Census recorded that 84 people lived within the Fort Wingate area. As an entity, they cultivated 51 acres and owned 2,177 sheep, 1,045 goats, 41 cattle, 101 horses and 13 mules or burros. Four of these households occupied log cabins; the remaining resided in traditional hooghans.[19]

In 1925, the Bureau of Indian Affairs (BIA) leased buildings at the old fort from the War Department and used them as boarding schools for Zuni and Navajo children. The entrance to Fort Wingate Administration building area had a large metal gate Navajos called *Beesh Daadilkal* ("iron door"). An estimated five hundred students were enrolled in boarding school by 1926, and Navajos continued to live and work in the area thereafter.[20]

In 1928, the status of Fort Wingate changed to Fort Wingate Ordnance Depot. In the months just prior to the Japanese attack on Pearl Harbor, December 7, 1941, Fort Wingate depot is said to have stored 46,000,000 pounds of explosives. In the 1950s, many Navajos worked at Fort Wingate Ordnance Depot. From 1950 to 1958, the depot maintained, modified and renovated stored ammunition during the Korean War. Munitions transported to the installation were disassembled using hot water to flush their contents. Wash water, operational ordnance and explosive waste were then pumped into storage and drying tanks. Overflow from the tanks drained into leaching beds and was deposited on soils and along watercourses.[21] In 1963, Pershing missiles were test-fired from White Sands Missile Range to Fort Wingate. By the mid-1960s, Fort Wingate stored as much as 14,622 tons of high-level explosives, which were then shipped to Vietnam.[22]

As computer technology advanced and Internet use became available in the early 1980s, aerial technology became more readily available, and advancements in satellite imagery technology compromised military base locations. Satellite imagery clearly shows bunker formations and military bombing test range sites, resulting in the closure of several ammunition depots formerly used as defense sites, bombing fields and military testing areas.

LAND REPATRIATION

After the decision was made to close the Fort Wingate Depot Activity (FWDA) army base, the city of Gallup lobbied for land tenure of the FWDA area, while both the Navajo and Zuni Nations sought sole possession of the former depot lands on the premise of "rightful heir." In 1993, the U.S. Department of Justice decided that the FWDA area lands would be returned to the "public" when the military had no more use for them. The Department of the Interior stipulated that the BIA would manage clean parcels of land for the Navajo and Zuni tribes, who were identified as the rightful heirs to the land; the remaining contaminated areas would be managed by the U.S. Army.[23]

More than eight hundred abandoned igloos (bunkers) now cover 60 percent of the area. Residents from neighboring communities may be exposed to explosive compounds and to rat droppings via skin contact with floors and walls of abandoned buildings or by the inhalation of particles of dust containing lead, rat saliva or droppings.[24] The Administration Area that included Building 5 has extensive amounts of pesticides in the soils; the extended storage yard, fire training grounds, former coal storage area, former paint shop and former storage yard all contain arsenic. The old trash burning ground contains beryllium, and the sewage treatment plant contains arsenic. Workshop areas containing ammunition painting/acid washout facilities and the former deactivation furnace contain arsenic. The former TNT Washout Facilities contains arsenic and explosives; all landfills including the current landfill contain arsenic.

Hazardous Contamination and Pathways of Exposure

Skin contact with contaminated sources of water, ingestion of contaminated fish and animals, drinking water and the inhalation of contaminated dust particles are pathways of exposure for surrounding communities. Pregnant women, young adults and children of this age cohort are most at risk to contamination due to the fact that children and young adults inherently breathe faster than do adults, as air is circulated through their bodies at a faster rate, thereby making them most susceptible to contamination. Children in contact with the ground are more likely to use sources of water for recreational purposes. Young adults are also more likely to use abandoned dirt roads for recreational driving. Since the time from conception to age twenty is the most formative and genetically sensitive period in human development, health risk prevention should be based on protecting the youngest generation, ages zero to ten, as well as women of child-bearing age, from contaminants.

Depending on the dose, frequency and duration of ingestion of agricultural products, as well as contaminated domestic well water, communities within and adjacent to Fort Wingate are at risk of developing adverse health effects. The populations of private land owners who currently allow livestock to graze on land west of the Fort Wingate Army Depot Activity area are at risk of migratory air and land pollution coming from the FWDA area. Both the Navajo and Zuni Nations consider contamination of water sources sacrilegious, so the protection of water sources at the FWDA area must be a priority. Immediate concern to both the Zuni and Navajo tribal nations is the threat of groundwater and surface water contamination. Prioritization of cleanup must consider the topography, direction of surface water flow, groundwater transport of chemicals and culturally significant areas in relation to one another.

The Church Rock Spill

Fort Wingate is located 8.2 miles east of the town of Church Rock. On July 16, 1978, at 5:00 a.m, just fourteen weeks after the accident at Three Mile Island, more than 1,100 tons of uranium mining waste-tailings gushed through a mud-packed dam near Church Rock into the Rio Puerco. With the exception of bomb tests, Church Rock is the largest nuclear spill to have occurred in the United States.[25]

The United Nuclear Corporation's dam wall was an earthen structure with a clay core, twenty-five-feet high and thirty feet wide. On the morning of the accident, a twenty-foot-wide section of it gave way, wreaking havoc downstream.[26] Even before the dam had been licensed, "the company's own consultant predicted that the soil under this dam was susceptible to extreme settling which was likely to cause [its] cracking and subsequent failure."[27]

Ironically, the Church Rock earthen dam was a "state-of-the-art" structure. Paul Robinson, an Albuquerque-based expert on mining issues, warned the Udall hearings that "UNC–Church Rock was the most recently built and the most carefully engineered tailings dam in the state." Similar dams owned by Anaconda, Kerr-McGee, UNC-Homestake Partners and Sohio, however, were also considered "disasters waiting to happen."[28]

Water from the spill traveled downstream from the Pipeline Arroyo along the Rio Puerco. By 8:00 a.m., water in the Rio Puerco in Gallup, New Mexico, nearly fifty miles downstream from the spill, measured six thousand times over the allowable standard of radioactivity.[29] The wall of water backed up sewers and lifted manhole covers in Gallup, twenty miles downstream, and caught people all along the river unawares. "There were no clouds, but all of a sudden the water came," remembered Herbert Morgan of Manuelito, New Mexico. "I was wondering where it came from. Not for a few days were we told."[30] Contaminated water continued its course along the river, crossing state borders into Arizona.

Uranium is extracted from the sandstone, usually by grinding it into a fine powder and then leaching it with sulfuric acid, which carries off the desired isotopes. However, waste from this extraction process contains 85 percent of the ore's original radioactivity and 99.9 percent of its original aqueous volume. Wastewater from the spill had a pH of less than 2 and a gross alpha particle activity of 128,000 picocuries per liter (pCi/1), leaving deposits of radioactive uranium, thorium, radium, polonium, dregs of metals such as cadmium, aluminum, magnesium, manganese, molybdenum, nickel, selenium, sodium, vanadium, zinc, iron, lead and high concentrations of sulfates, in soils seventy miles downstream.[31]

In the book *Ecocide of Native America*, authors Donald Grinde and Bruce Johansen state,

> *The River is the main source of water for the area. Very few newspapers carried the story and those that did stated that since the area was "sparsely populated" there was no "immediate health hazard." Health officials issued press releases warning people not to use the water. Many of the people in the area could not read English and didn't have electricity, radios or television sets. Livestock sickened and died. Workers were sent to the area with buckets and shovels to clean contaminated soils for several weeks, however only 50 tons of the tailings were cleaned up.*[32]

A New Mexico demographics report of 2015 shows that a population of 1,051 lived in Church Rock. The median household income was $24,938; 40 percent of residents lived in poverty. An estimated 93.9 percent were of American Indian racial/ethnic groups, followed by 2.6 percent Hispanic and 1.3 percent Asian.[33] Most non-Indians living in the Church Rock Chapter are married to Navajo. The majority of the inhabitants of the area speak Navajo. The primary livelihood for the chapter's residents is livestock, supplemented by "traditional" self-employment, which includes jewelry making, silversmithing, sewing, stone carving, wood carving and weaving.[34]

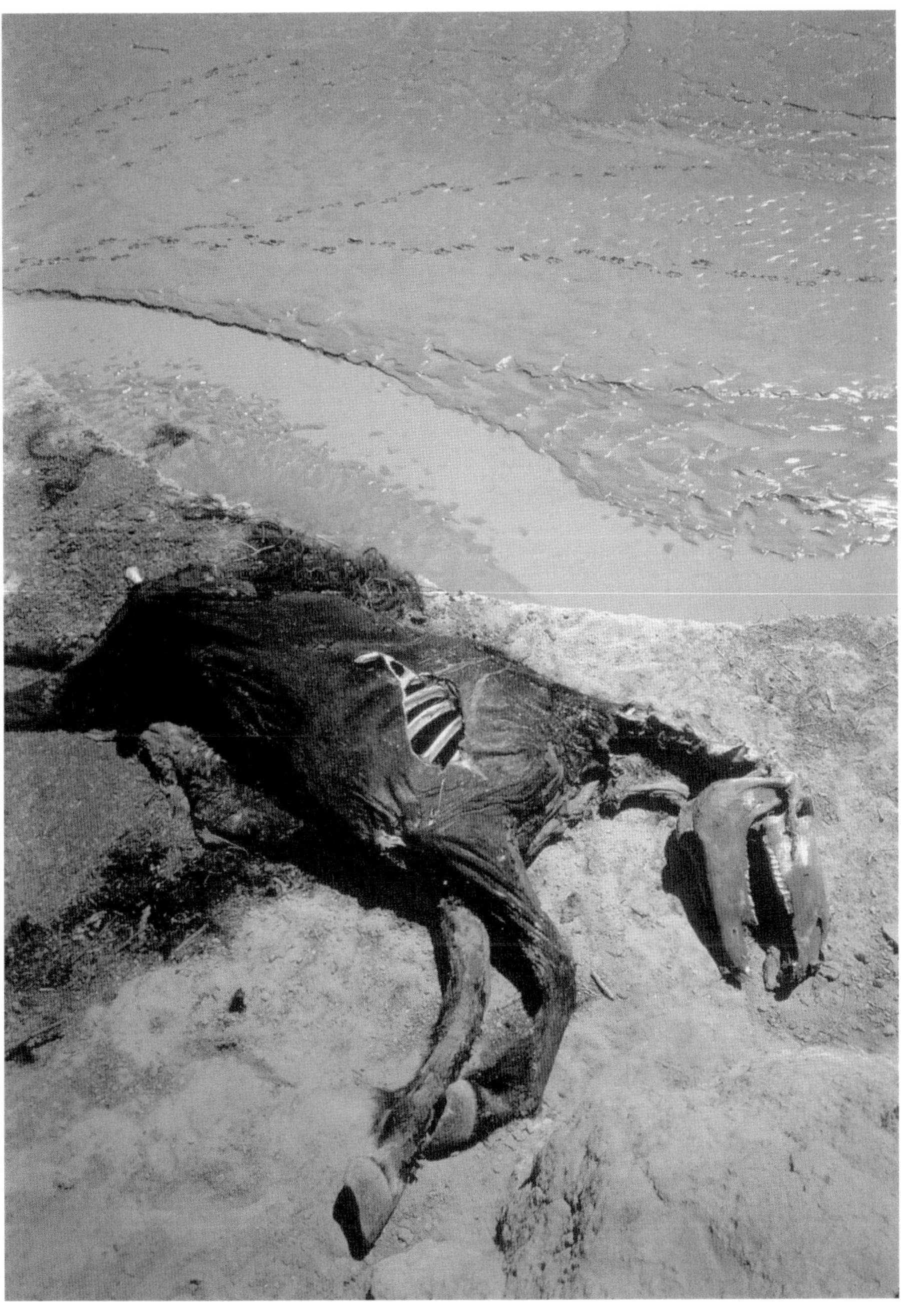

This horse died from drinking water contaminated by the nuclear spill into the Rio Puerco, 1979. *Photograph by Dan Budnick. ©1979 Dan Budnik. All Rights Reserved.*

Cleanup

"We have removed more than 3,500 tons of potentially affected sediment from the streambed to a distance of more than 10 miles from the mill," J. David Hann told the Udall hearings. "The combinations of these clean-up efforts, and natural effects, such as rain, have largely restored normal conditions in the area."[35] Robinson pointed out that UNC had in fact removed just 1 percent of the tailings and liquid known to have spilled from the dam. More than eighteen months after the accident, indications were strong that radiation and other pollutants had penetrated thirty feet into the earth. A report by a Cincinnati-based firm brought in as a consultant by the EPA warned that at least two nearby aquifers had been put "at risk."[36] An Arizona water-quality official complained in an interview that the rains had merely transported the pollutants into his state.[37]

Uranium contamination lasts such a long time that disposal practices must consider groundwater preservation: "The stark contrast between a typical 20-year mill life and an 80,000-year half life for the dominant radionuclide (thorium 230) necessitates a much greater forward look than is now evident in waste disposal practices and preservation of ground-water quality." As early as 1964, the Federal Water Pollution Control Administration told a congressional hearing that fish caught downriver from the Naturita and Uravan uranium mills showed higher radium concentrations than those caught upriver. Downriver hay samples also showed contamination, as did cows' milk. "In this case," said the authorities, "the prime source of radium intake for the cows is believed to be from eating hay irrigated with contaminated river water."[38]

A study of eleven animals by the Centers for Disease Control confirmed animal contamination and warned that kidneys and livers of local livestock might contain high doses and should not be eaten. The CDC also warned locals not to drink water from the river and to avoid its banks during windstorms, when radioactive particles might be more easily inhaled. The CDC emphasized that radiation levels in local animals did not exceed New Mexico standards, but it cautioned against consumption, as "the health risks of low doses of radiation" were "not completely understood."[39]

Dr. Carl Johnson, director of Colorado's Jefferson County Health Department, further warned that detectable radiation levels in the tissues of children might surface only "over a period of many years." UNC had promised to provide local residents and their animals with clean drinking water, but an Arizona newspaper confirmed that the company was delivering

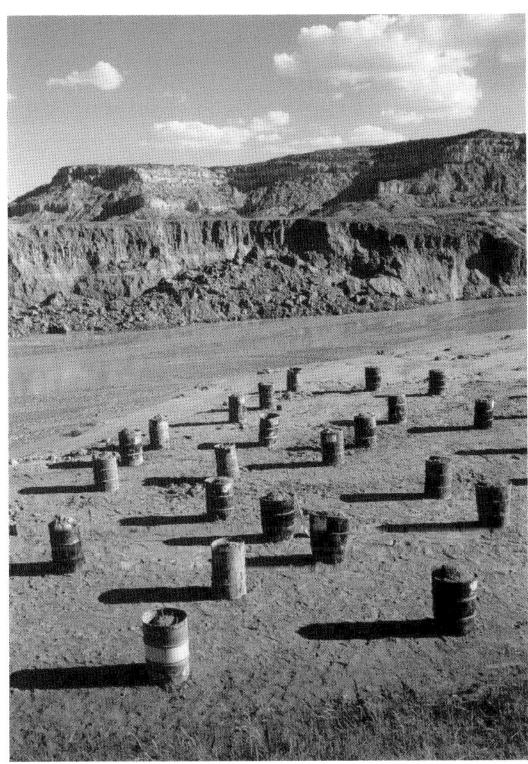

Left: Cleanup of a nuclear uranium waste spill along the Rio Puerco, on the Navajo Nation, 1979. *Photograph by Dan Budnick. ©1979 Dan Budnik. All Rights Reserved.*

Below: Jimmy Arviso, a Navajo sheepherder tending his flock near the Rio Puerco River, near Church Rock, New Mexico, November 2, 1979. *Photo credit: Center for Southwest Research, University Libraries, University of New Mexico.*

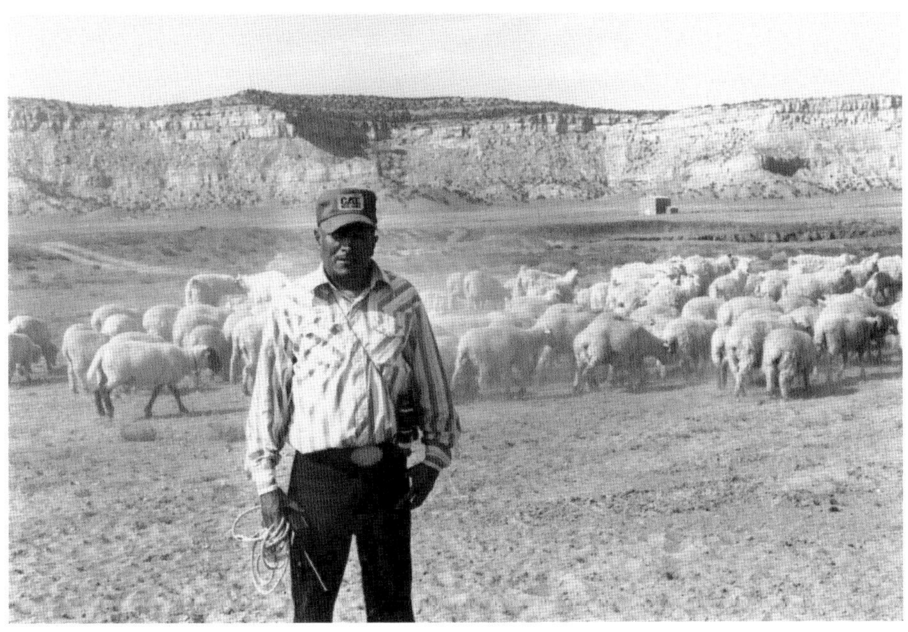

just half the promised amounts. A request by some of the downstream residents for emergency food stamps to replace their lost livestock was also denied by the government.[40]

A resident from Church Rock explained: "The UNC had put up signs saying 'contaminated wash, keep out.' But our cows, sheep and horses can't read that. Most of us can't read, write, or speak English. The signs do no good. If [neighbors] know we are from the Rio Puerco wash, they won't shake our hands," he added. "They think we have a high level of radiation. They ran from me. They are afraid of us. That's why people look at us, that's why no one comes to help us. It is wet now, but on days when it dries up, the wind will come along. The dust settles on the grass. The sheep eat it. We eat the sheep. We wonder what that does to our lives."[41]

Health authorities warned Navajos against eating their sheep because they might be radioactive from drinking the Rio Puerco, which was contaminated by an accidental spill of 100 million gallons of radioactive water from a uranium mine. At the time of the photo on November 2, 1979, Navajo sheepherder Jimmy Arviso said, "Doctor says danger. I don't know," explaining he understands he is not supposed to allow his sheep to drink from the river or eat mutton, but he's not sure why or how serious it is."

The Church Rock Spill received far less media coverage than that of Three Mile Island. Many scholars suggest this is an example of environmental racism, which led to purposeful neglect from the U.S. government regarding the health, compensation and contamination cleanup in affected communities. Due to the contamination and mining legacy at Church Rock and Fort Wingate Depot, all future land use of the area has been altered indefinitely, and the land can no longer be lived on or used without the consideration of its toxic nature. While financial incentives for uranium mining exist, strong opposition remains. The Navajo Nation issued a uranium mining ban in 2005 and continues to hold its stance while many surrounding communities and environmental groups persist, voicing their resistance to further uranium mining as well. Residents in and around the town of Church Rock continue to live in a polluted environment; many suffer from health issues, tumors and high incidences of cancer.

2
FOUR CORNERS

Land Sacrifice Zone

It is no accident, therefore, that the "remote," "sparsely populated," West has been targeted so often in recent years by eastern policy makers as the receptacle of deadly technological experiments. It is not just that we get these experiments because there's more room out here. It's more deliberate than that. This is a consciously evolved policy, a dangerous trend in United States history, and it has resulted in a de facto national policy of trading ecosystems—national sacrifice areas—for economic gain, scientific data and national security.
—Grinde & Johansen[42]

Uranium Mining

When Spanish colonists arrived in New Mexico, they found silver, gold, copper and turquoise mines already in operation. Lured by the vast mica sandstone "sparkly rocks" of the region, these colonists thought the state was wealthy in silver and gold. Currently, New Mexico is one of the U.S. leaders in the outsourcing of uranium and potassium salts and also has prolific mines of petroleum, natural gas, coal, copper, gold, silver, zinc, lead and molybdenum.[43]

An article written in 1993 by Timothy Benally Sr., director of the Office of Navajo Uranium Workers, gives a brief history of uranium mining on the reservation and in the Four Corners area. Benally states,

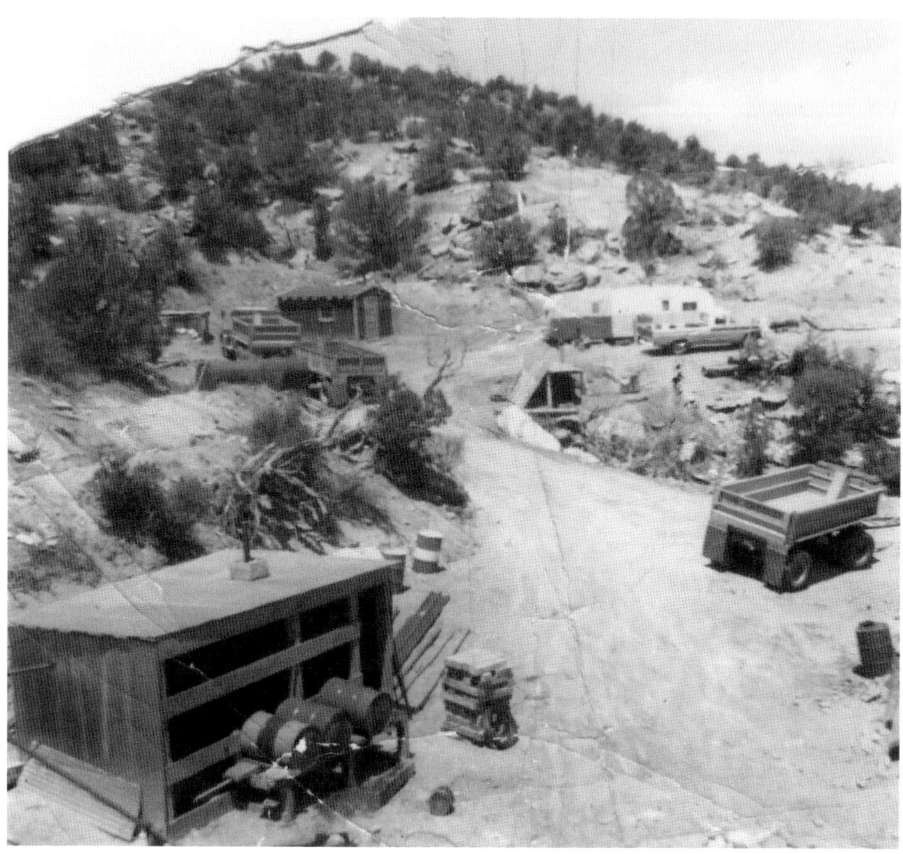

Uranium mine near Cove, Arizona. *Photo credit: Flora Garnanez.*

> *Mining started around 1918 in the Carrizo Mountains when companies were mining vanadium. Uranium was an accidental discovery. At the time of discovery, the miners and processors didn't know exactly what it was that they found, so they just kept it around often in gunnysacks. Around World War II, the government realized the resource and its potential, and soon prospecting and digging all over the area for uranium. By the 1950s, the Vanadium Corporation of America and Kerr-McGee were the principal mine owners. Jobs were scarce on the reservation and Navajo men, particularly for men who had just come back from the armed services, and so many men jumped at the chance to work in the mines, even though they were paid at a rate below what was acceptable in the rest of the country.*[44]

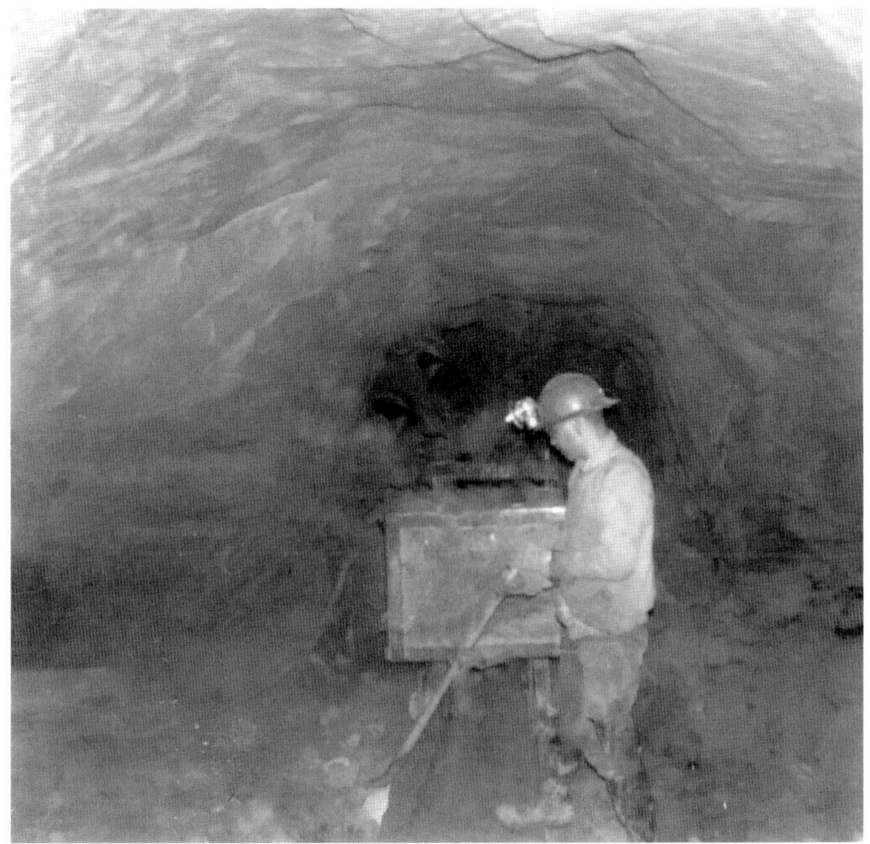

Navajo miner James Garnanez. *Photo credit: Flora Garnanez.*

Mining companies claimed to know nothing of the dangers involved with the mining and processing of uranium, but previous scientific experiments showed that there was a danger to humans. Benally felt that the miners were used as human guinea pigs and that the result of mining uranium was far worse than he could have ever imagined. Uranium mining would leave the land poisoned and harm the health of New Mexicans for many generations to come. Benally said, "Uranium was stored in houses, hooghans and trading posts. Dirt and rock from the mine tailings were used in home building and repairs, to pave roads and parking lots, and to cover playgrounds. Until very recently, children played on the tailing piles near abandoned mines. There are still many unsealed mines on the reservation and uncovered tailings. Whenever the wind blows, which is more often than not in this country, radioactive dust becomes airborne."[45]

Black Mesa

Demand for coal between 1879 and 1882 augmented aggressive mining and production of natural resources in New Mexico. As the railroad system arrived, the primary exportation industry was of coal and lumber to California. Starting in the 1880s, in what is considered a sacred mountain to the Dine', heavy logging began in the Chuska Mountains. During the 1960s, logging operations continued under the Navajo Forest Products Industries (NFPI) enterprise, accumulating nearly $20 million in debt and decimating forest areas.[46]

Black Mesa, known as a sacred mountain for Navajos, had a coal strip-mining operation on it owned by Peabody. In 1970, the new technique of "coal slurrying," which is the combination of crushing coal rocks and mixing them with water to transport via pipeline, became cheaper than the process of underground mining, which required heavy machinery. Thus, coal coming from Black Mesa was combusted, becoming electricity that is transmitted across the Navajo reservation to major cities like Los Angeles, Phoenix and Tucson. In order to meet the energy demands of the nation, the United States told the Navajos that there was an energy crisis and per capita consumption of electricity had doubled from 1963 to 1975. Navajo land was leased at almost three times as much as that of the Crows and Northern Cheyenne during this time of negation, and only half of Navajo homes had electricity.[47]

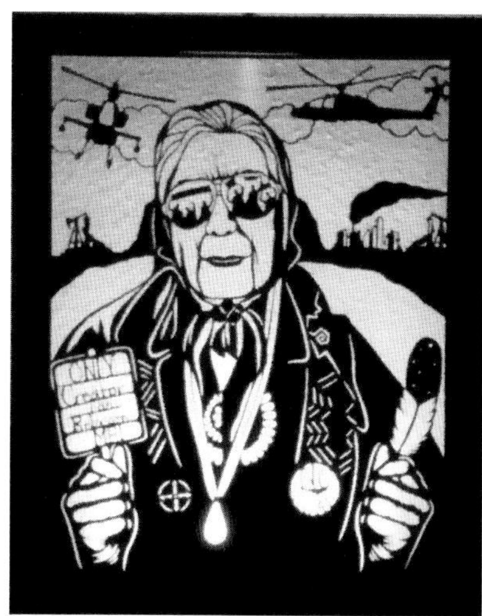

Papercut image of Roberta Blackgoat. *Valerie Rangel.*

Traditional Navajos voiced their opposition to coal and uranium leasing, construction of electricity and gasification plants and landscape destruction while environmental pollution to energy development mounted. Increasing ethical debates and conflict disputes over land rights ensued. "To many Navajos, digging coal is sacrilegious and a form of energy colonialism. To them, the coal rush is only a transmutation of the gold rushes and land rushes which drove Indians off their lands. It is an old story; the developers want the resources, and the Indians are in the way."[48]

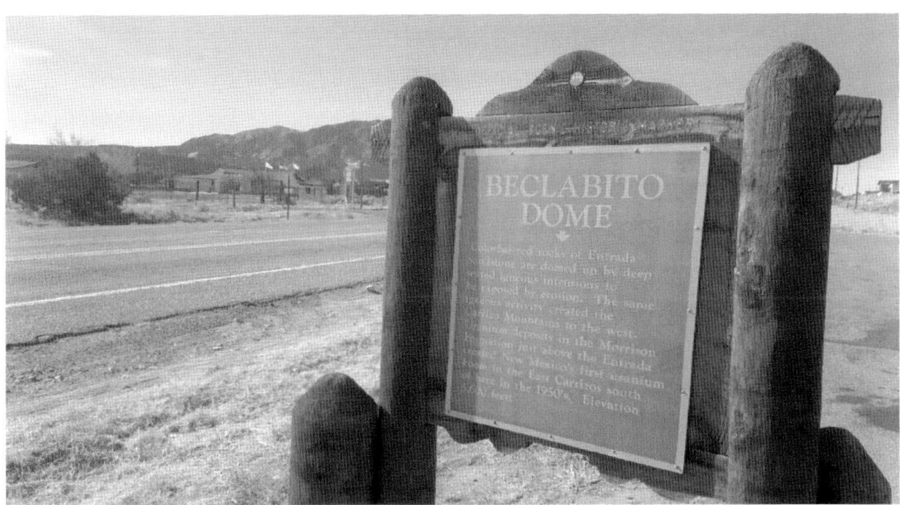

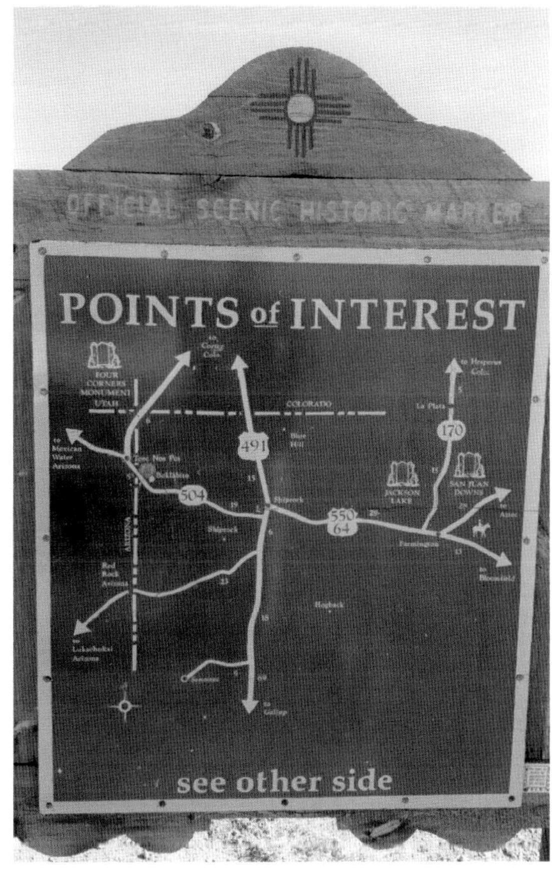

This page: Official New Mexico State Scenic Historic Marker (*Above*: front; *right*: back). *Photo credit: Earl Tulley*.

Land Sacrifice Zone

Fred Johnson advocated for land reclamation that involved more than filling holes in the land and covering them with grass. "True reclamation," he said, "would restore the ecosystem of plants and animals." In a report from the National Academy of Sciences, ecosystem restoration is considered an impossible feat in areas that receive less than ten inches of rainfall a year. The report suggests that development in areas with such sensitive ecosystems would have to be deemed "National Sacrifice Areas."[49]

The Four Corners area would be called a "Land Sacrifice Zone." In the early 1960s, the Four Corners Power Plant was built near Shiprock to take advantage of the coal resources in the area. During this time, grassroots organizing began advocating for the health of surrounding communities with concern for the environmental effects of coal and uranium production and extraction. Early protestors were often silenced through bribes, threats and violence. Members of the American Indian Movement joined traditional and environmental activists, and the Coalition for Navajo Liberation (protesting uranium) and other Navajo groups formed.[50] In 1974, Congress passed Public Law 93-531, creating a Relocation Commission, which forcibly relocated ten thousand Navajos and one hundred Hopis out of their homelands for a Peabody Coal strip mine. Those who opposed or resisted relocation faced such antics as livestock seizures, fencing by government, a housing construction ban and other types of harassment.[51]

Cost to Human Life

According to Greenpeace:

> *When people who work in mines, or live close by them, inhale coal dust and carbon, this hardens their lungs, leading to black lung disease (also called pneumoconiosis or CWP). People living near coal mines have higher-than-normal rates of cardiopulmonary disease, chronic obstructive pulmonary disease, hypertension, lung disease, and kidney disease. Local communities also suffer when coal fires occur. These fires emit toxic levels of arsenic, fluorine, mercury and selenium, contaminants that can enter the air and food chain of local communities.*[52]

Dr. Lora Shields conducted a study of Navajos in the Four Corners region. She reviewed 13,329 births at the Indian Hospital between 1964 and 1981 and found cases of cleft palate, club feet, and Down syndrome, with a high rate of stillbirths and infant deaths occurring before 1975.[53]

The Four Corners Power Plant is the nation's largest source of nitrogen oxides. In 2011, the plant emitted 14,488,178 tons of carbon into the air, 24 million pounds of sulfur dioxide, 77 million pounds of nitrogen oxides, 8 million pounds of soot and 1,297 pounds of mercury. The American Lung Association estimates that sixteen thousand people in the region (15 percent of the population) suffer from lung disease presumably caused by plant emissions. Coal combustion waste from the mines supporting the Four Corners and San Juan plants contaminated local groundwater with sulfates, which led to the death of livestock. According to Earth Justice's 2012 report, 70 million tons of coal waste (containing cadmium, selenium, arsenic and lead) were dumped in the Navajo Mine and 80 million tons in the San Juan mine.[54]

The ecological cost of coal production is damaging and often irreversible. Strip mining is used to excavate coal that is buried near the surface. This destroys ecosystems by removing topsoil that contains plants, minerals and microbes. Mining companies may completely remove portions of a mountainside or mesa to reach deeper pockets of coal, thereby altering the geomorphology of the area. Denuded landscapes lead to gullying, erosion, landslides and flooding, as well as less annual precipitation due to the lack of trees to draw moisture from clouds. Water runoff from denuded areas has the potential to contaminate agricultural land within the region.

Coal Power Generation

Coal slurrying and power generation drains underground aquifers and contaminates surface waters by releasing heavy sediment loads into streams and river channels; this can choke fish and aquatic ecosystems. Once coal is crushed and mixed, water containing heavy metals starts to settle out and acidic water is created. This orange-colored acid mine drainage can accumulate within abandoned mines and persist for decades or centuries after a mine closes. Any contact with such acid drainage would render surface water unsuitable for human consumption and kill animal organisms and aquatic life.

The Central Arizona Project (CAP) canal was designed to deliver, on average, 1.5 million acre-feet of water annually (equivalent to 1.5 million football fields one foot deep) from the Colorado River diversion point at Lake Havasu to the Phoenix and Tucson regions. CAP operations began in the '80s with the help of the Navajo Nation. Arizona would not have been able to sustain its growth and development without CAP.[55]

Navajo coal mined on the Black Mesa traveled by train to the Navajo Generating Station near Page, Arizona, where it burned to create steam from Navajo water taken out of the Colorado River at Lake Powell; steam generated electricity at the plant. The volume of Navajo water used annually in the plant is about 33,000 acre-feet. During the past thirty-plus years, the total Navajo water used is about 1,000,000 acre-feet. At today's lease prices, that is over $1 billion worth of water, for which the Navajo Nation has never received compensation, even though the surrounding states have acknowledged the water used is part of the Navajos' share of the Colorado River. The Page Plant sends electricity through power lines across the reservation and on to other states, with 20 percent being used for CAP to pump Colorado River water 336 miles uphill to central and southern Arizona.[56]

On February 13, 2017, the Navajo Nation, the Salt River Project, Arizona Public Service, Tucson Electric Power and Nevada Power Company, as well as one of the largest customers, Central Arizona Project, determined that operating the Navajo Generating Plant (NGS) was unsustainable. All parties agreed to operate through December 22, 2019, at which time the plant would be decommissioned, which would include the removal of the plant buildings, water facility, transmission and railroad starting on July 2, 2017. Complete restoration of the Ash Landfill, evaporation ponds, solid waste landfill, the asbestos landfill and perched water for thirty years would take place by December 22, 2020.[57]

The New Mexico Public Regulation Commission agency will ultimately decide on PNM's approval request for abandonment of the San Juan Power Generating Station, and this will also directly affect the adjacent mine, and the decision process could be settled by parties without community consultation or input. New Energy Economy Inc. and Con Alma are focusing a health impact assessment on the concerns and vision of affected communities, which will help to inform the decision-making process. Communities residing on Navajo Nation lands in New Mexico are presently living without running water or electricity. Injustice takes place when resources are extracted and "dirty coal" power

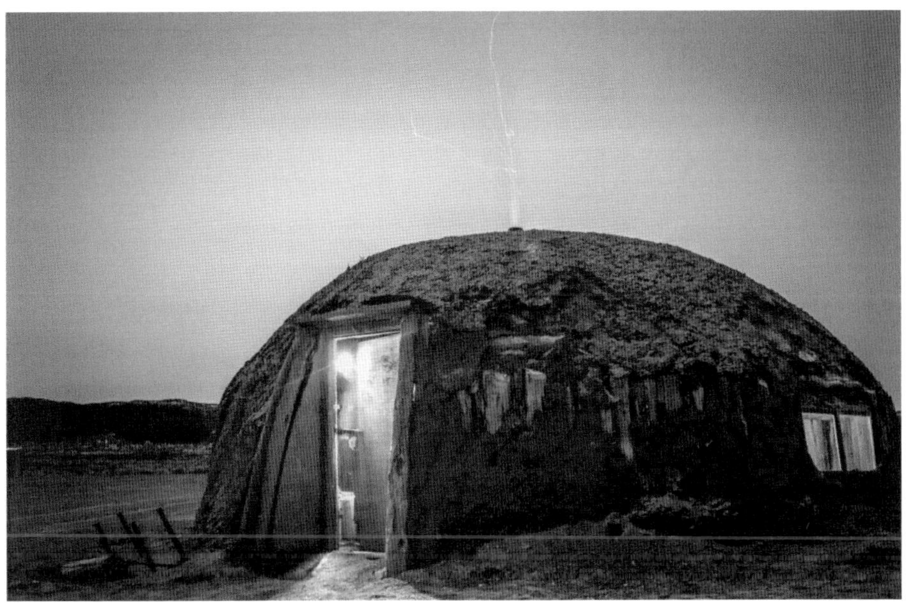

Traditional Navajo hooghan with battery-powered light, no electricity. *Photo credit: Daniel Lombardi.*

generation occurs in impoverished communities that do not benefit from the electricity produced. *High Country News* explained this phenomenon: "Social economists refer to this kind of thing as colonialism and even economic racism."[58]

Groundwater Concerns

Coal power generation includes a coal slurry that transports sludge to a power generation gasification plant, where more water is consumed. A coal gasification plant requires an estimated 10,000 acre-feet of water a year. The Black Mesa mine's last day of operation was December 31, 2005. It used groundwater from the Navajo Aquifer to transport coal 273 miles to the Mohave Generating Station (MGS) in southern Nevada. The process required 4,600 acre-feet, or 1.3 billion gallons, of pristine groundwater from the Navajo Aquifer every year for coal transportation.[59]

The National Institute of Science calls the Navajo Aquifer one of the most pristine water sources in the nation, and it is one of the few sources of drinking water in the United States that naturally meets the Environmental Protection Agency's standards for drinking water. According to the October 2000 Natural Resources Defense Council report, the aquifer showed serious declines in water level, and water levels have decreased more than one hundred feet each year in some wells on Black Mesa since pumping for slurry ensued.[60]

Local residents are concerned about plans to expand mining operations, specifically the health effects of mining and burning of coal and the effect on employment and tribal revenues. There are also concerns about how much the mining has already shifted a cultural and spiritual understanding of the earth and the protection of the water itself. Communities around Black Mesa fear that due to the drawdown from long-term and mass quantities of pumping, water quality and their future water supply are at risk should mining operations ensue.[61] Several other coal-fired power plants—Navajo, Escalante, Springerville, Coronado and Cholla—may be affecting air and water quality in the San Juan Basin. The Coronado and Cholla plants are owned by the Salt River Project and serve the residents of Phoenix and surrounding areas.

Having failed to install court-mandated pollution control improvements at a cost of $1 billion, and without a means to transport or a facility to receive its coal, the Peabody Western Coal Company (PWCC) was forced to shut down the Black Mesa mine in January 2006. In an effort to keep the mine open and operational, the PWCC and the Navajo and Hopi tribes explored ways to transport coal, open a power plant on Black Mesa or convert the coal into diesel fuel. Alternative energy production such as wind and solar are also being considered by the Navajo Nation. Residents and environmental groups continue to voice concerns for any future planning and development, as well as issues of repatriation of artifacts and bones dug up by the Peabody coal mine.[62]

> *The West is a colony, and change will not come from the outside. Until the West, as a region, develops strong mechanisms to protect itself along with locally based, self-sustaining economies, it will continue to be raped and plundered like any other colony.*[63]

For more than four decades, coal and water extraction has been a point of contention sparking controversial debates within the Four Corners region of New Mexico. Residents, local grassroots organizations, Navajo traditionalists and concerned environmental groups have sought to retain their homelands and protect the land from environmental destruction. Immediate impacts from the mine closure were economic; 75 percent of the Hopi tribe's annual income was generated from the Black Mesa mine, which also employed many Navajo and Hopi tribal members. Communities are concerned about economic justice, energy justice, environmental restoration and the possibility that coal mining operations resume in the decades to come.

Within the Four Corners region, there are profound places of natural beauty and of archaeological and cultural significance, such as the Grand Canyon, Arches National Park, Canyonlands, the Petrified Forest, Mesa Verde and Chaco Canyon. These protected places face threats from aggressive mining of resources such as uranium, oil, gas, coal oil shale and tar sands. Environmental damage is evident in the physical landscape and includes scars by nuclear weapons testing, accidental spills, nuclear waste tailing spills and seepage from leach ponds, coal strip mining and abandoned mineral mines dotting the landscape.

3

NAVAJO WATER HAULERS

Water is fundamental for a life of human dignity. It is a prerequisite to the realization of all other human rights.
—World Health Organization

The introduction to a documentary called *The Water Haulers*, originally broadcast on New Mexico PBS station KNME, begins with the narrator saying, "Water is essential to every living thing. All those, need a continuous supply of water in order to live. A desert is a sign of meager rainfall. Life here is greatly restricted."[64]

According to a 2004 study conducted by the Navajo Nation EPA, an estimated seventy thousand people, approximately 30 percent of Navajo Nation residents, haul their drinking water because they do not have piped water to their homes. This population of Navajos living without running water is comparable to the population of Santa Fe, New Mexico's capital city. Though the Navajo Nation has advised against the use of unregulated water sources for drinking water, many of the "water haulers" choose these nearby sources rather than driving long distances to haul their water.[65]

Often, water haulers rely on nonpotable water sources such as stock tanks for their drinking water supply. Those who do have running water depend on public water supply systems that are deteriorating; many have exceeded the maximum withdrawal capacity of their source aquifer, have poor water quality and are susceptible to drought. Lack of reliable and affordable

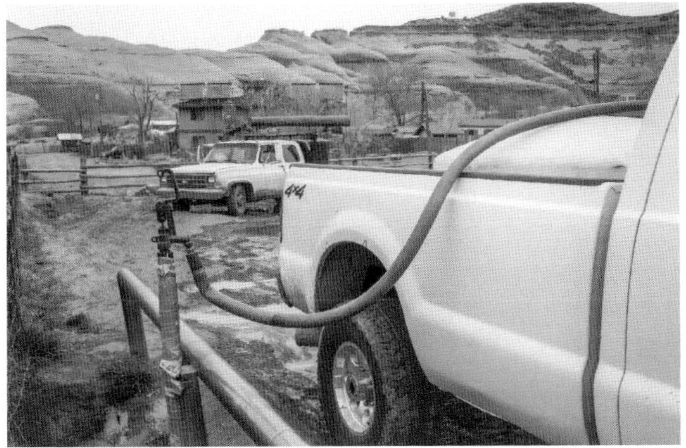

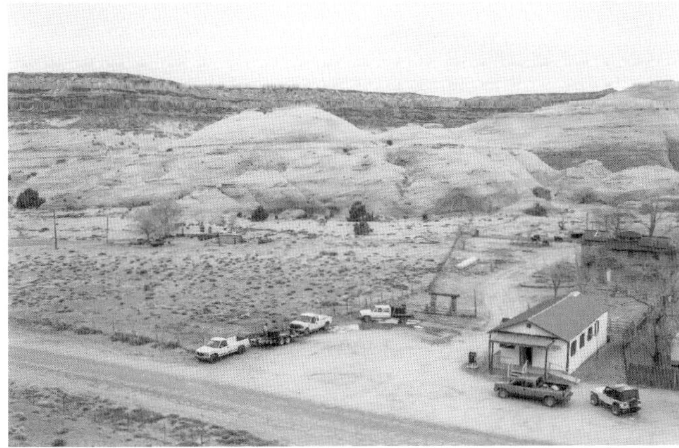

Top: Will Yazzie's tank slowly fills in the back of his truck. This is the only way his family has access to clean drinking water. It takes about an hour each way to drive to the water tap at the center of the community; sometimes, waiting for your turn in line can take even longer. *Photo credit: Daniel Lombardi.*

Bottom: Most of the residents of Monument Valley have no functioning tap water at their homes and must use a community tap to fill tanks. *Photo credit: Daniel Lombardi.*

potable water supply stifles economic growth throughout the reservation and contributes to a high incidence of disease and infection attributable to waterborne contaminants. Inadequate and unsafe water supplies burden the federal financial system and health-care programs as well as contribute to higher mortality rates. Lack of access to safe water and in-home sanitation facilities has led to the emigration of 21.6 percent of the population between 1990 and 2000 from the reservation.[66]

In 2018, Navajos in New Mexico living without running water are unable to tap nearby river basins because they do not own the water rights, cisterns are too expensive to dig into the hard bedrock, water delivery by truck is not available to homes and the majority of wells are contaminated from uranium mining in the region. Navajo water haulers must transport all of their water for domestic household purposes such as cooking; washing dishes, fruits and vegetables; bathing children; and feeding livestock. Often, Navajos will set buckets outside to capture rain and runoff from rooftops. Chemical compounds from roofing material may be the cause or contribution to waterborne illnesses such as skin rashes, abdominal pain and dysentery that many Navajo water haulers experience.[67]

A Huffington Post article featured the story of water hauler Shirley Peaches, who resides in Tall Mountain in the Navajo Nation, twenty-five miles from the nearest paved road and functioning water faucet. Peaches says, "My family still uses melted snow to wash dishes and wash clothes." Drinking water is hauled from Shonto, a reservation town in northern Arizona. It's a two-hour drive round-trip when the road is passable. In February, Peaches says, it was dicey: "The snow is melting; the road is muddy. You have to get there early in the morning when the ground is frozen, and you can't get back until 11 p.m. when the ground is refrozen."[68]

In stark contrast to neighboring urban centers such as Gallup and Phoenix, Navajos without running water limit their consumption to around ten gallons a day per person, utilizing outhouses instead of sewage systems. By comparison, a low-flow showerhead uses two gallons per minute, and washing dishes efficiently by hand uses at least eight gallons.[69]

In October 2009, the Centers for Disease Control and Prevention (CDC), the Navajo Nation Environmental Protection Agency (NNEPA) and the Diné Network for Environmental Health (DiNEH) Project, in partnership with the

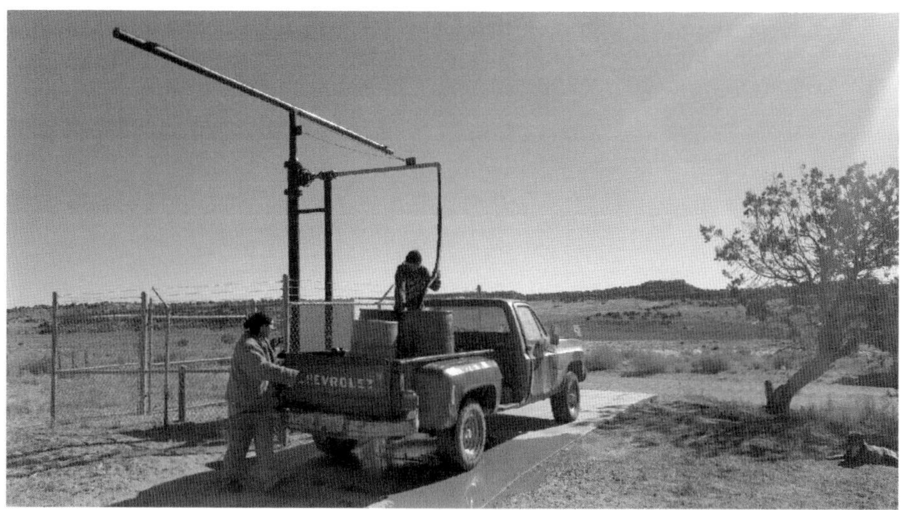

Rincon Marquez water station, Whitehorse Lake Chapter. *Photo credit: Earl Tulley.*

U.S. Environmental Protection Agency Region 9 (USEPA), sampled thirty-six unregulated water sources on the Navajo Nation in the Eastern Agency and chapters in the South and North Central areas of the reservation. Sampling of wells on the eastern Navajo Nation was intended to assess unregulated water sources near abandoned uranium mines for Superfund Site Screens in the Eastern Agency and Superfund Preliminary Assessments for Marino Lake and Moonlight Mines, as well as other potential public health threats; only twelve water sources exceeded primary drinking water standards.[70]

WATER JUSTICE

The Navajo Nation asserts that it has had "senior water rights" since the signed Treaty of 1868, which effectively ended internment at Bosque Redondo and created what would be the present Navajo reservation lands. Though the treaty did not specify water rights, in a 1990s New Mexico court case, specific and exact language written in the 1849 Navajo Treaty was also found in an 1852 water rights case involving the Mescalero Apache tribe, and the court ruled in favor of the Indian water rights. Possessing "senior water rights" would allow the Navajo Nation "first in line–first in right" authority to use the San Juan River water, first and in any way that they choose. According to the 2004 water rights settlement, however, Navajo water rights were restricted and could not interfere with other water rights users, effectively granting them only a little more than half of the water in the San Juan River basin.[71]

Other tribes throughout the nation have water rights issues stemming from undefined treaties and the delineated area of land grants still in dispute. In 1908, a Supreme Court decision, *Winters v. United States*, declared that Indian reservations came with implied rights to the amount of water "sufficient to fulfill the reservation's purpose." The Indian Relocation Act was supposed to provide incentives to indigenous communities to relocate tribal members from Indian reservations to urban centers. This effort was an attempt to move indigenous peoples from a system of federal dependency to one of self-autonomy and assimilation into the dominant society. The court, however, could not define how much water tribes were entitled to, and the Winters Doctrine could not be enforced in the case of Navajo Water Rights.[72]

According to water right laws, one must "use it or lose it," meaning that, as river water passes through a person's land or private property, and if they choose not to use it for agricultural or domestic purposes, the water right is passed on to the next downstream water rights user. For Indian lands, water would not be lost if they chose not to utilize it, however, in areas where indigenous communities reside, Arizona and Utah for example where poverty is rampant and water use is low, the government has discriminatorily ruled in favor of non-Indian water projects that had the capital to implement infrastructure for irrigation.[73]

In 2004, a water rights settlement between the Navajo Nation and the State of New Mexico was reached. According to the final agreement, the Navajo Nation maintained only 56 percent of its San Juan water rights in exchange for a massive infrastructure project that promised to bring running water to parts of the reservation that have gone without it for centuries and, in turn, secure water reserves for future population growth and economic development opportunities for the state.[74]

The Navajo Nation's water rights settlement failed to effectively negotiate with the adjoining states of Utah and Arizona so that all water managers could share claims to the San Juan River basin equitably. Arizona withdrew from negotiations, and Utah wanted the Navajo Nation to agree to go without water in times of drought, even though power plants, cities, golf courses and industry of mass urban centers would be furnished with water first, holding precedent before the water needs of the tribe.

Ambrose Yanito, a Navajo Nation resident, explains his view of water rights. Gesturing across the plain to distant cities, Ambrose says, "Everybody has got a swimming pool out there, just about. If the tribe had more of its water, people could grow vegetable gardens and eat healthier, perhaps reducing the 22 percent rate of diabetes on the rez. We can't grow nothing out here because those people over there are just jumping in the water and wasting it. What would happen if we turned that off? How would they feel? We gave them their life! See?"[75]

Water as a Human Right

Is access to water a basic human right? Is this issue an ethical debate?

On July 28, 2010, through Resolution 64/292, the United Nations General Assembly explicitly recognized "the human right to water and sanitation" and acknowledged that clean drinking water and sanitation are essential to the realization of all human rights. The resolution calls on states and international organizations to provide financial resources and help in the capacity of building technology and infrastructure to help countries, in particular developing countries, provide safe, clean, accessible and affordable drinking water and sanitation for *all humans*.[76]

According to the World Health Organization (WHO),

> *Between 50 and 100 litres of water per person per day is needed to ensure that the most basic needs are met. A sufficient water supply is defined as a quantity that is sufficient and continuous for the purposes of: drinking, personal sanitation, washing of clothes, food preparation, personal and household hygiene. Usually determined by national and/or local standards, drinking-water quality or "safe water," must be clean source of water, acceptable in color, odor and taste, and is free from micro-organisms, chemical substances, and radiological hazards that constitute a threat to a person's health.*[77]

The UN further describes water security as possessing water facilities and services that are culturally appropriate and physically accessible within or in the immediate vicinity of the household, educational institution, workplace or health institution. According to the WHO, "the water source has to be within 1,000 metres of the home and collection time should not exceed 30 minutes." Costs associated with water facilities and services for a community must be affordable for all. The United Nations Development Programme (UNDP) suggests that water costs should not exceed 3 percent of household income.[78]

THE PROMISED GALLUP-PIPELINE PROJECT

The Navajo Nation is a sovereign government within the United States, but is it considered a developing country? Since water availability is connected to economic growth, surely investment in water security would serve in the best interest of preventing civil unrest, legal battles and ongoing threats to public health and safety that burden medical providers and add to human suffering.

Since New Mexico was granted statehood in 1912, an estimated 30 percent of the Navajo Nation's enrolled members continue to live without access to clean, safe, sufficient, accessible and affordable sources of water, a clear violation of the UN's declaration of basic human rights. Today, thousands of Navajos suffer from waterborne illnesses and commute long distances to haul water in a region with few economic opportunities. Water is vital to the pursuit of freedom and happiness and is most definitely connected to economic development and capitalistic growth. Why, then, are thousands of Navajos denied this basic human right and necessity?

The UN's World Water Development Report for 2016 found that three of four jobs globally are either heavily or moderately dependent on water and that water shortages and lack of access to water and sanitation could limit economic growth and job creation in the coming decades. "Water and jobs are inextricably linked on various levels, whether we look at them from an economic, environmental, or social perspective," said the director-general of UNESCO, Irina Bokova, in a statement.[79]

In 2009, Senators Jeff Bingaman and Tom Udall and Representative Ben Ray Luján signed the Record of Decision for the Navajo-Gallup Water Supply Project Planning Report and Final Environmental Impact Statement. The project is key to the Navajo Nation–San Juan River Basin Water Rights Settlement. The proposed project would provide municipal and industrial water supply to the eastern section of the Navajo Nation, southwestern part of the Jicarilla Apache Nation and the city of Gallup, New Mexico; diverting 37,376 acre-feet of water from the San Juan River and an estimated 250,000 people. The project would include 260 miles of pipeline, 24 pumping plants and 2 water treatment plants in place and operational by the year 2040.[80]

Secretary Salazar said, "This project addresses an unfulfilled promise to support the Navajo people by providing a long-term sustainable water supply that will reduce the need for hauling water, improve health conditions on the Reservation, and provide the foundation for future economic development activity in northwestern New Mexico."[81]

Senator Bingaman, who chairs the Senate Energy and Natural Resources Committee and sponsored the water settlement legislation in the Senate, commented, "Today's action means we can move ahead with this important pipeline project, finally bringing water to thousands of Navajos who are currently not served and bringing water certainty to Gallup."[82]

Senator Udall remarked, "This settlement addresses the basic need for water and sanitation that I believe is the right of every individual, and which is long overdue in this region of New Mexico. The agreement that led to this settlement is the culmination of years of work by many parties. Today we celebrate the completion of the final EIS for the Navajo Gallup

Water Supply Project, the next important step in providing a dependable water supply for the Navajo Nation, the Jicarilla Apache Nation and the city of Gallup."[83]

Representative Luján also stated, "Water availability is a critical issue in New Mexico. Many tribal communities on the Navajo Nation do not have access to a relievable water supply, and the Navajo-Gallup Water Supply Project will provide many of these communities with stable and reliable access to water."[84]

The Navajo Nation has entered into water settlement negotiations with each neighboring state. In 2010, a deal with New Mexico funded a fourteen-year construction project that will deliver and serve water to many Navajos for the first time. Arizona's deal with the Navajo Nation disintegrated in 2012, and Utah's settlement with the Navajo Nation Council passed and must still be approved and funding provided by the U.S. Congress.[85]

Since the $1 billion New Mexico water settlement deal was completed, the Bureau of Reclamation built two water lines, two water treatment plants and several pumping stations; infrastructure for the first water line runs one hundred miles along the New Mexico border with Arizona. Water from the San Juan River travels nearly four hundred miles from southwestern Colorado to Lake Powell, to users as far as fifteen miles south of Gallup. The second water line will run along the eastern portion of the reservation in central New Mexico. It is important to note that even after the Gallup-Pipeline project has been completed, not all Navajos in New Mexico will have running water due to some houses being too remote to pipe infrastructure.

The entire Gallup Pipeline project is slated to be complete by 2024. In the meantime, some tribal communities such as Whitehorse Lake have started to benefit from running water in their homes as a result of the water rights settlement. During the time that Whitehorse Lake had no access to running water or electricity, the population declined and residents relocated to border towns. Now that Whitehorse Lake has obtained water, people from the community who had relocated are moving back. The town plans to build police and fire stations, a convenience store, a gas station, a road maintenance yard for heavy equipment and an increase in housing development.[86]

The Navajo Nation is the largest Indian reservation with the largest population; its culture is vibrant and its language still spoken and written today. Younger generations relocated from tribal lands to pursue western education and employment opportunities within urban centers, as well as to

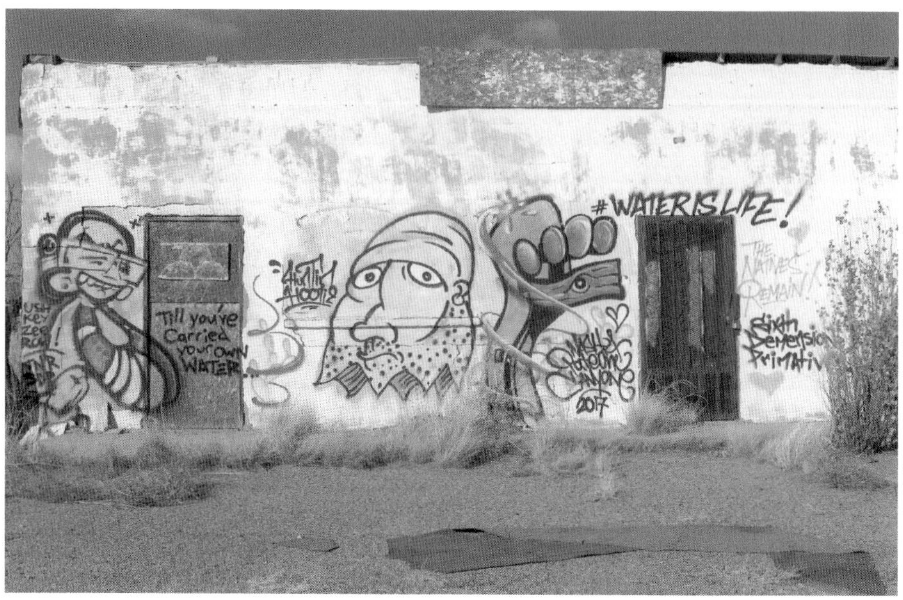

Street art in Counselor, New Mexico. *Photo credit: EvalynBemisPhotography.com.*

have access to better food and health care in places where water is readily available. For Navajos choosing to live in their homelands, and for those who practice traditional farming and sheep herding, life is difficult—hauling water; dealing with economic insecurity and contaminated lands and water sources and health issues; and living without running water in homes that often lack electricity in remote regions. It is important to acknowledge the physical, historical, spiritual and cultural connections that Navajos possess to these homelands and why many choose to persevere with resolve in order to endure such harsh conditions.

4
FUTURE OF URANIUM

Tso Dził

American Indians hold their lands-places as having the highest possible meaning, and all their statements are made with this reference point in mind.
—Vine Deloria Jr., God Is Red

Located in New Mexico's San Mateo Mountains, Mount Taylor is a mountain peak that is visible from up to one hundred miles away with an elevation of nearly twelve thousand feet. The mountain has been a pilgrimage site for as many as thirty Native American tribes; many consider this place "holy" and have incorporated it into their traditional cultural and religious ceremonies.[87] Before the mountain was named after President Zachary Taylor, it was known to the Acoma as Kaweshtima ("place of snow"), to the Hopi as Tsiipiya, to the Zuni as Dwankwi Kyabachu Yalanne and to the Navajo Nation as Tso Dził. Tso Dził, the turquoise mountain, is one of the four sacred mountains marking the cardinal directions and the boundaries of Navajo traditional homelands. The mountain holds special significance for the Pueblo of Acoma. "Mount Taylor is the mountain that is associated with the cardinal direction of north, which for us here at the Pueblo of Acoma is the direction for which all things began," said Theresa Pasqual, director for the pueblo's Historic Preservation Office. "There are a number of deities associated with that mountain."[88]

Uranium ore is present on Mount Taylor, which is owned by federal, state and private parties. Although the New Mexico Mining and Minerals Division has received proposals for exploration, mining and milling operations on Mount Taylor, there has been opposition to exploration permits and concern for the cultural and natural resources of the area by tribal nations and environmental groups. Mining would affect the landscape itself and contaminate the primary water source for the Pueblo of Acoma. Surrounding communities have endured decades of open-pit uranium mining in the areas around Laguna and Acoma and the town of Grants. Exposure to uranium caused sickness and death; these communities are dealing with diseases associated with tailings and toxic water discharges, as well as a legacy of unmonitored open-hole pits that are susceptible to winds and surface water connections that further contaminate the region.[89]

The Mining Law of 1872 emerged as the result of the California Gold Rush and other mining booms in the 1800s. Since mineral deposits were found predominantly on federal lands, laws needed to be created in order to transfer mineral rights from public ownership to private mine ownership. The legislation that was created allowed U.S. citizens and companies the right to mine for minerals and establish rights to federal lands without authorization from any government agency. Under the Mining Law of 1872, once a claim has been made by a private land owner or a company, it is difficult for federal agencies to deny the mining permit even if there are other public uses such as recreation, or environmental concerns such as wilderness protection or preservation of natural and cultural resources.[90]

Most of the uranium mined in the United States has been taken from the Colorado Plateau region, with more than two hundred mines located in eighteen New Mexico counties, producing a large share of the total U.S. output. When uranium prices dropped in the 1980s, many mines and mills closed, leaving behind uncontrolled waste and other public health and safety hazards, as well as contaminated water supplies. Though some uranium mill sites have undergone cleanup by the federal government, land and water restoration is still needed to protect adjacent communities, particularly tribal communities that have been severely impacted.[91]

In an area twenty-five miles north of Grants, New Mexico, near Ambrosia Lake, are three of New Mexico's five licensed uranium mills, two of which have been abandoned. An estimated 82 million tons of radioactive uranium mill tailings are stored in piles at the mill sites. In 1979, Grants was called the "Uranium Capital of the World," housing eight thousand workers. Today,

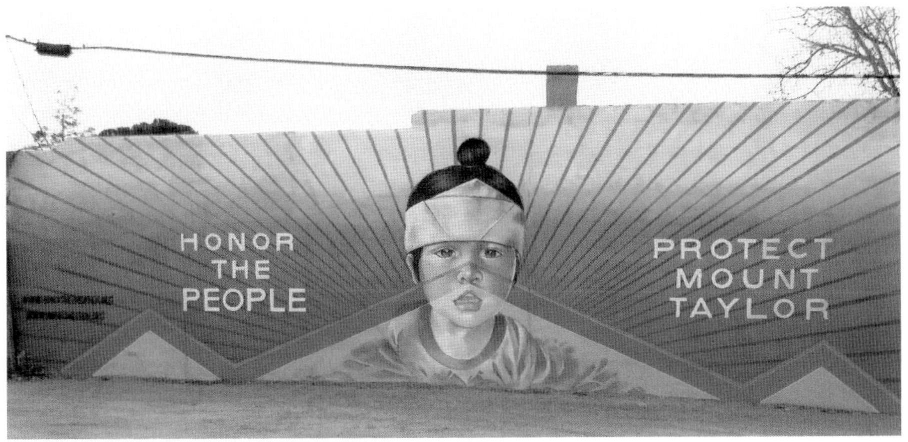

Protect Mount Taylor mural. *Photo credit and art by Nani Chacon.*

there are dozens of abandoned mines in New Mexico, with fewer than five hundred workers employed at active mines.

From 1979 to 1982 and from 1985 to 1989, an underground mine formerly owned by Chevron produced uranium ore on Mount Taylor. Due to the low cost of uranium on the market, the mine was shut down in 1990, and the Rio Grande Resources, which had been a subsidiary, bought the mine in 1991.[92] For the past twenty-five years, the mine had been inactive; it had been on "standby" status for the last fifteen years. Mines on "standby" status aren't required to remediate their pollution or the surrounding environment, and therefore Mount Taylor has not had pollution cleanup. Mining renewal applications have been filed to reactivate and resume mining processes. Many concerned individuals from the Navajo Nation and nearby pueblos stood in opposition to reactivating the mines at a hearing in Grants on December 4, 2016.[93]

PROTECTION

Federal and state agencies possess the power to continue issuing mining permits on small parcels of land without consulting tribes. New Mexico's tribal communities grew very concerned over a permit that was issued without notification at the site of "a reburial pursuant to the Native American Graves Protection and Repatriation Act." In June 2007, the All

Indian Pueblo Council passed a resolution calling for the protection of sacred sites on Mount Taylor. Then, all nineteen pueblos and the Navajo Nation petitioned the state historic preservation officer for an emergency traditional cultural property status.[94]

The tribes assert that allowing exploratory mining and drilling on Mount Taylor would create the need for more roads, which would further desecrate the sacred peak. They also asserted that the exploratory sites themselves would disturb sensitive soil and various plant species used for medicinal and cultural purposes. In 2009, the National Trust for Historic Preservation listed Mount Taylor as one of the ten most endangered historic sites in America.

In 2009, the Pueblo of Acoma led the pursuit of Mount Taylor becoming a "Cultural Property Listing," on the premise that the pueblo people have lived with this sacred site since "time immemorial." Although tribes of the region do not own any parcels of land on the mountain, there are significant numbers of archaeological sites associated with early Chaco people, including religious shrines and a small village.[95]

After the Cultural Properties Review Committee's announcement, Navajo Nation president Ben Shelly said that the designation, called "permanent," meant that cultural resources would be protected: more than 686 square miles of land that includes Mount Taylor and nearby mesas would be protected. However, according to the National Trust for Historic Preservation, it isn't clear what the designation means for existing activities, but it does limit uranium mining in the area. The cultural properties designation states that tribes must be consulted prior to the issuance of permits, but it does not grant veto power over permit decisions.[96]

Shortly after the cultural property designation, a group of landowners and uranium mining companies sued the state cultural agency and the tribes, claiming that the area was too large to be protected as a historic site and that even minor developments were hindered that were seen as a violation of their property rights. In February 2011, the permanent designation was overthrown by the Fifth Judicial District Court. Private property land ownership occupies one-fourth of the area, and even if the private owners "opt-out" of the cultural designation, the mountains' historical landscape is said to be maintained and activities such as hiking, biking, snowshoeing and snowmobiling on National Forest Service lands will be allowed to continue.[97]

Mount Taylor remains a sacred site to many tribes within the region. Although Mount Taylor is currently listed as a Cultural Property, which means that the tribes must be consulted prior to any issuance of drilling permits, the designation does not mean that mining operations will cease indefinitely.

Surrounding communities endured decades of open-pit uranium mining, with Laguna and Acoma Pueblos and the town of Grants experiencing high rates of diseases associated with tailings, toxic water discharges and a legacy of unmonitored open-hole pits, waste piles subject to winds and contaminated surface waters. Further destruction of the sacred mountain would only create more radioactive air pollutants and water quality issues from mining uranium, thus contributing to an increase in health disparities for New Mexicans.

LOS ALAMOS

Most of northern New Mexico was a rural agrarian society with ranching communities containing Hispanic and indigenous populations until the 1940s and World War II.[98] German advancements in technology put pressure on the U.S. government to develop atomic power, leading to a secret program of nuclear testing and development codenamed the "Manhattan Project." By the end of 1941, the federal government's Office of Scientific Research and Development, headed by scientist Vannevar Bush, took control of the project. The U.S. Army Corps of Engineers was tasked with building plants, laboratories and research and testing facilities.[99]

The Manhattan Project was essentially a program based on developing man's most destructive weapon: the atomic bomb. In order to create such a weapon of mass destruction, two key ingredients were necessary: plutonium and uranium. At Oak Ridge, a medium-sized reactor for uranium-235 and plutonium was built; this facility produced the majority of uranium used to build the "Little Boy" bomb, which was deployed over the Japanese city of Hiroshima in August 1945.

According to the City of Los Alamos's webpage, in 1942, a boys' ranch school became the headquarters of the Manhattan Project. By early 1945, shipments of enriched plutonium from the Handford plant in Washington State, which had three reactors, were being sent to Los Alamos every five days. This enriched plutonium material would be used to test the first atomic bomb. Led by J. Robert Oppenheimer, scientists detonated a plutonium bomb at a test site located on the U.S. Air Force base at Alamogordo, New Mexico, at 5:30 a.m. on the morning of July 16, 1945.

Before the testing of the atomic bomb, the rural population surrounding Los Alamos as well as the Trinity Site in New Mexico lacked running water

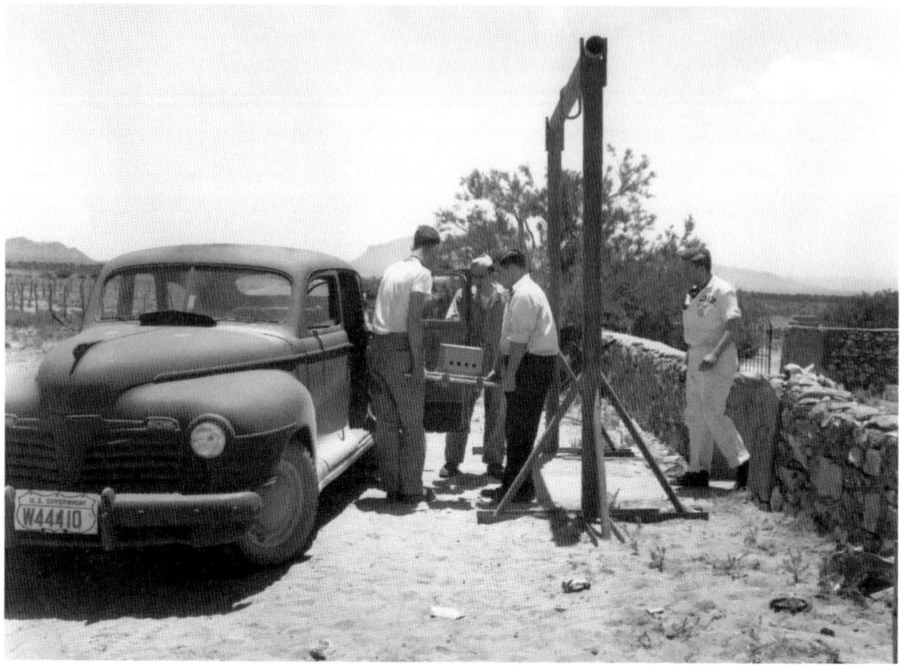

Workers at Los Alamos National Laboratory transport equipment for nuclear testing at Trinity Site, New Mexico. *Photo credit: Palace of the Governors.*

and relied on cisterns and holding ponds for their water supply. Residents of neighboring communities were never warned or evacuated prior to the testing, nor were they informed of potential public health hazards associated with the atomic testing fallout.[100]

The town of Tularosa, sixty-five miles south and downwind of the Trinity Site, had a population of around three thousand. At the time of the bomb test, witnesses reported that "radiation levels near some houses were almost 10,000 times above the level currently allowed in public areas." A group called the Tularosa Downwinders believes that the Trinity test is responsible for causing cancer and other illnesses in residents of communities surrounding the atomic test site.[101]

Cofounder of the Tularosa Downwinders, Tina Cordova, said, "The group decided to educate the thousands of visitors coming to the opening of the Trinity Site. We decided that it would be a good opportunity for us to peacefully demonstrate to basically inform people of the fact that the bomb actually damaged people's health. They were exposed to radiation and their health has been damaged because of that." Cordova said, "The government

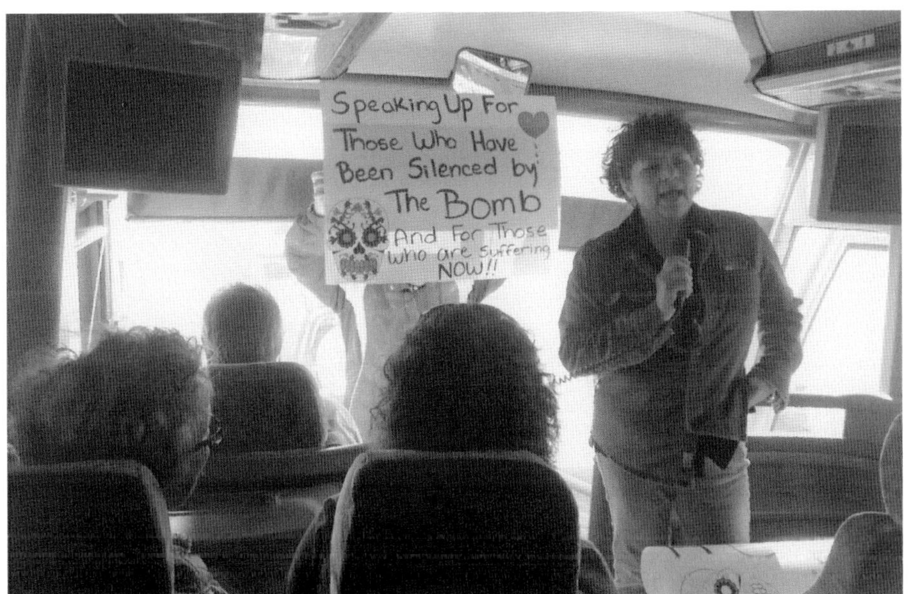

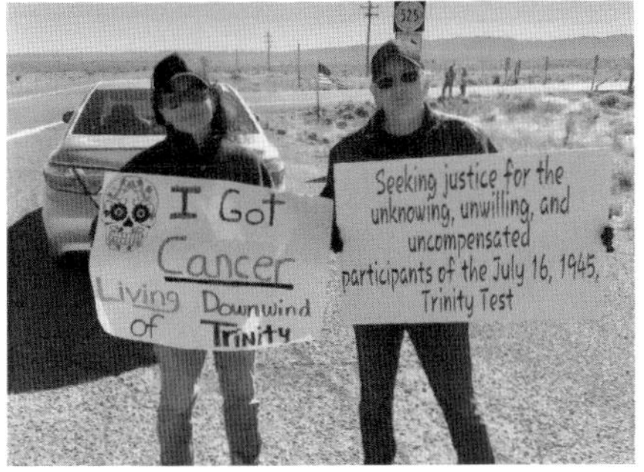

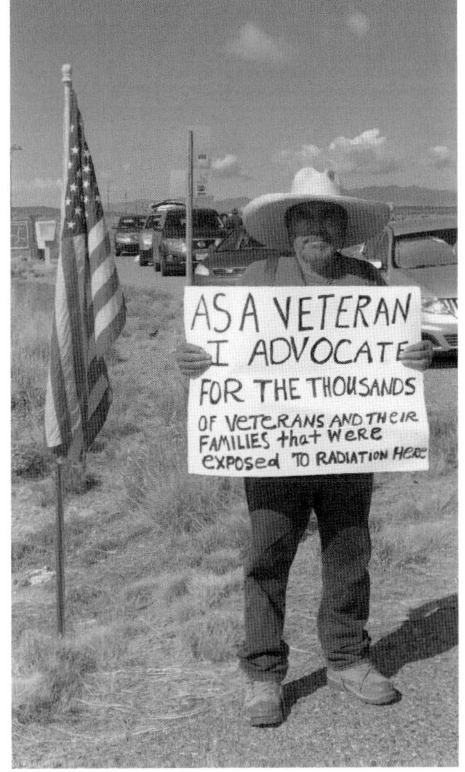

This page: Tularosa Basin Downwinders Consortium demonstration at the Trinity Site, New Mexico, October 1, 2016. *Photo credit: Tina Cordova.*

has always characterized it as no one was living in the area but in reality, within a 60 mile radius, there were probably 40 or 50,000 people that lived there at the time. That's not remote and uninhabited, that's a lot of people and all those people were exposed to radiation."[102]

CONTAMINATION

The fork of what is known as Acid Canyon was the main dumping point for Los Alamos National Lab from 1940 through the '60s. Precipitation and snowmelt runoff carried contaminants downstream and infiltrated the soils, seeped into cracks and penetrated groundwater basins along drainage pathways. In 2000, an Associated Press article stated, "Radioactive contaminants and other chemicals have been found in storm water…the worst of contaminated material seems to be coming from the forest above the lab. Cyanide found in the runoff draining from the lab was thought to originate from fire retardant used during the Cerro Grande fire, however other elements such as cesium-137, strontium-90, and plutonium were likely to have come from fallout during the 1950s and 60s."[103]

As the Cerro Grande Fire swept across the Jemez Mountains, it charred an old dump site called Material Disposal Area R at Los Alamos, and although the materials did not explode, they ignited and burned underground for more than a month. Concerned state officials said the World War II–era dump probably contained high explosives, depleted uranium, barium, beryllium and heavy metals. While the Cerro Grande fire raged, it released radioactive and hazardous airborne contaminants from Los Alamos National Laboratory (LANL) property and from burning vegetation and debris in the mountains in which it charred. The fire left significant damage to the environment, denuding steep slopes of nearby canyons and mountainous areas, greatly increasing the speed at which surface water runoff travels. During the annual monsoon season, high-intensity flash thunderstorms, soil erosion and the potential for mudslides or rockslides also increased. Precipitation falling on a wildfire-impacted watershed cannot easily filtrate into the ground because the loss of vegetation cannot slow runoff flow. Instead, high-intensity fires often cause soils to become hydrophobic, therefore repelling precipitation. This does not allow water infiltration into the ground, resulting in further surface runoff and erosion.

According to the New Mexico Environment Department:

> *Gross alpha activity increases in stormwater runoff are strongly correlated with the amount of suspended sediment carried in stormwater. Post-fire flash floods contain extremely high levels of suspended sediment and subsequently possess very high levels of gross alpha activity. Following the Cerro Grande Fire in 2000 the concentrations of several metals (e.g., copper, aluminum, barium, manganese, and zinc) increased and in some cases exceeded state water quality criteria. As the forest and soils recovered, these concentrations decreased and by 2010 these waters no longer exceeded state water quality criteria.*[104]

Where Does Nuclear Waste Go?

As nuclear scientists hurried to create the first atomic bomb, nuclear research, weapons development and testing created waste sites, filled containers and saturated the vegetation and soils where wastewater was dumped. Stormwater that contacted contaminated areas and waste piles carried radioactive materials downstream and contaminated water penetrated into the aquifer below.

The problem of waste containment and management has plagued LANL for decades. The EPA identified more than twelve hundred potential dump sites scattered around forty square miles of lab property on the Pajarito Plateau and authorized the New Mexico Environment Department to expand its hazardous waste program at the lab under federal law. The lab built a treatment plant to handle low-level radioactive waste, but in 1994, a plume of explosive chemicals was found leaking into the aquifer directly below the plant.

In 1999, the Waste Isolation Pilot Plant (WIPP) underground storage facility, located thirty miles from Carlsbad, opened. Radioactive waste materials would be stored underground in a deep salt cavern. The New Mexico Environment Department began overseeing the permit under which the lab manages its radioactive waste and ships it to WIPP. By 2002, the state was in court, battling the federal government and LANL's operators over cleaning up legacy waste and the waste generated by ongoing nuclear research.

The "Trinity Site" was chosen as a test site due to its rural nature, being "sparsely populated" and far from major cities. Unfortunately for

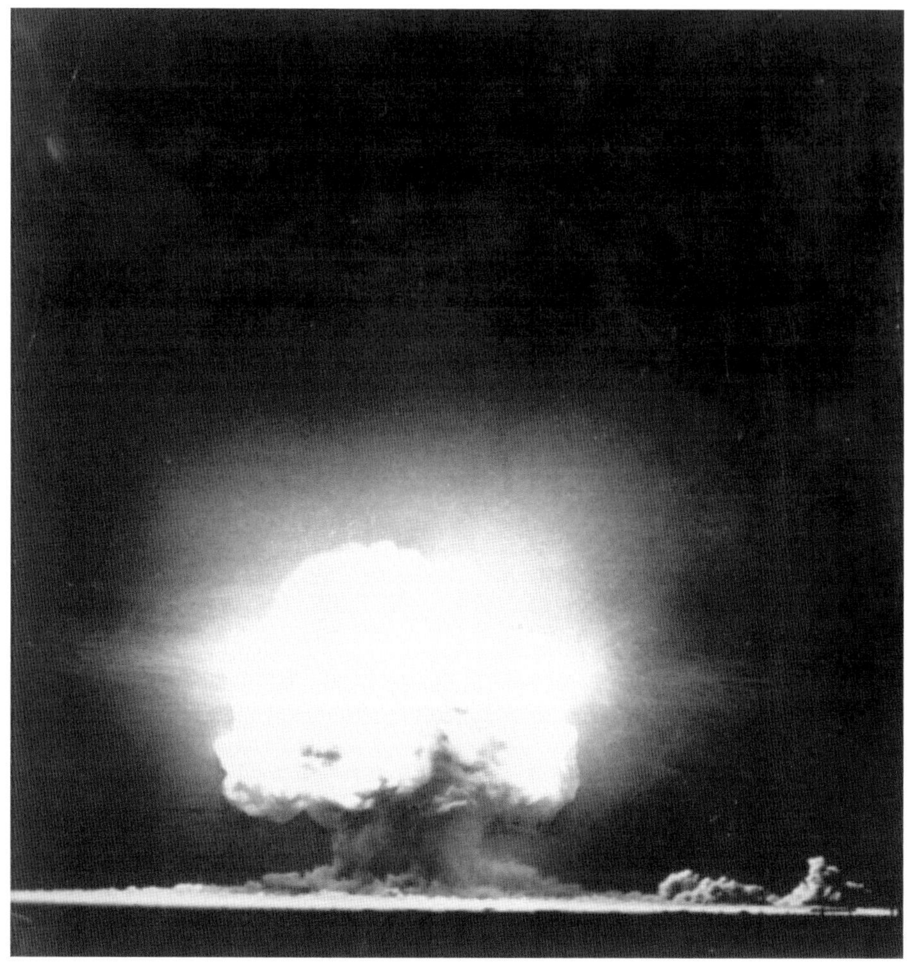

First atomic bomb detonated on July 16, 1945, at the Trinity Site, New Mexico. *Photo credit: Center for Southwest Research, University Libraries, University of New Mexico.*

the surrounding Hispanic and multi-tribal communities downwind and downstream, the testing of the first atomic bomb came as a great shock. This secret test not only frightened neighboring communities, it also caused suffering and illnesses related to the ensuing radiation fallout. Decades after weapons testing, downwinder protest groups are trying to educate the public about the truth of the "necessary discovery": that there were significant communities living in the area of Los Alamos and the Trinity Site at the time of atomic testing and that the health of these communities has been affected.

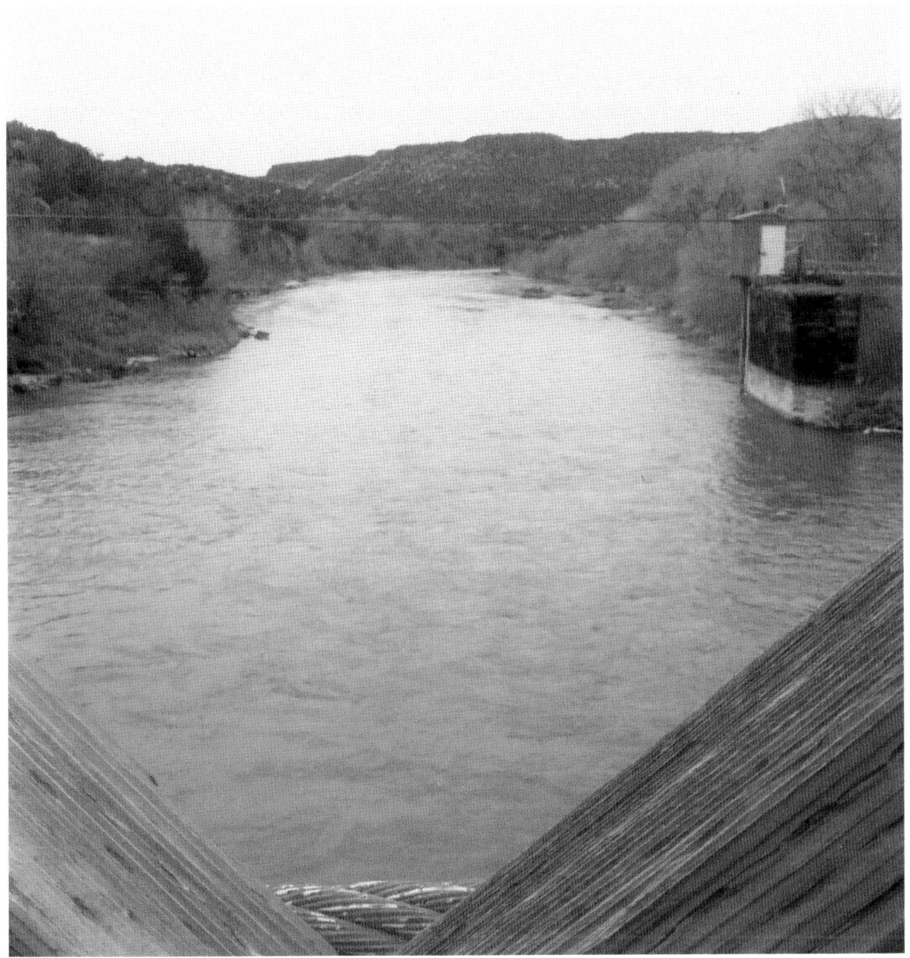

Rio Grande River at San Ildefonso Pueblo. *Photo credit: Valerie Rangel.*

Inevitably, another wildfire will char the areas of Los Alamos and spread contamination throughout the region, again releasing radioactive particles stored in vegetation. Meanwhile, community and environmental groups continue to pressure the Department of Energy and the current contractor/owner of LANL to clean up present and ongoing waste and legacy waste sites and to manage its stormwater more effectively. Concerned citizens as well as the New Mexico Environment Department and nuclear watchdogs raised concerns about the waste piles and their potential to contaminate the Rio Grande and groundwater.

Surrounding communities of Los Alamos were never notified about potential risks of exposure and never compensated for the health disparities caused by this "sacrifice zone." More important, these communities continue to experience health and safety risks due to the lack of cleanup at the lab's waste areas and the continued nuclear research that generates waste piles and contaminant discharges into New Mexico water. Due to interconnected hydrologic connections of the region, radioactive waste dump sites situated along the Rio Grande and its tributaries have allowed infiltration into the state's groundwater reservoirs. The contamination of underground aquifers has likely occurred since toxic dumping began in the 1940s. Unsuspecting well-water users and the municipality of Santa Fe now drink plutonium-laced drinking water. The City of Albuquerque may need to start planning for the treatment of radioactive particles from nuclear waste contaminants as it garnishes water from the Rio Grande to dilute arsenic levels as well as other sources of contamination.

THE FUTURE OF WIPP

The Waste Isolation Pilot Plant (WIPP) is an underground disposal facility or dump for plutonium-contaminated waste generated by the research and manufacturing of U.S. nuclear weapons. The facility dates back to 1957, when the National Academy of Sciences recommended bedded salt formations for long-term underground disposal of radioactive waste. WIPP started with a mission to "start clean, stay clean," with the goal of not opening waste containers, in order to limit the likelihood of radioactive and toxic chemical emissions that endanger workers and the public.[105]

Unfortunately, on February 14, 2014, one or more uranium storage drums leaked, releasing radioactivity into the environment and contaminating more than 8,000 feet of underground tunnels, including the exhaust shaft that goes 2,150 feet to the surface. A reported twenty-two workers on the surface were contaminated. Due to the contamination incident, workers now need to wear protective equipment with respirators. The Department of Energy admitted that the mine cannot be adequately maintained and that it has contamination from cracked drums with inadequate ventilation, as well as slabs of salt weighing several tons that fell from the mine ceilings in at least four areas. Before the WIPP facility is reopened, an analysis of six hundred potentially explosive waste drums emplaced underground must occur.[106]

Hydrogeological Connections

The WIPP site is located in an area of karst topography—holes caused by dissolution of soluble rocks such as limestone, dolomite and gypsum bedrock. Rainfall averages fourteen inches per year, with sand-covered surface areas that have high infiltration transmissivity rates. There are large sinkholes with windblown depressions in the topography. One of the most pertinent environmental concerns at the WIPP site is whether surface soils have hydrological connections to the karst geology below the site. If an accidental spill were to occur within the salt formations and cavernous regions of the WIPP site, would seepage infiltrate into the aquifer below and endanger the surrounding communities' water supply?

According to Brett Madres, in a 2011 report, "Storage and Disposal of Nuclear Waste":

> *The United States is currently the world leader in electricity generation from nuclear energy with its 104 reactors being the global high for a single country. Nuclear energy has a minimal contribution to greenhouse gas emissions and global warming while providing consistent energy, day and night, rain or shine, for the lifespan of the nuclear power plant. This consistency carries tremendous value that renewable energy systems like photovoltaic arrays and wind farms cannot match. An unanticipated consequence of the U.S.'s successful nuclear power program has been the accumulation of spent nuclear fuel that sits on site, in storage, all around the nation. Even though controversy involving high level waste always surrounds nuclear energy programs, nuclear energy will be needed by many countries for the foreseeable future.*[107]

Some radioactive waste generates heat and decay, while others form byproducts from spent fuel that contain half-lives of thousands to millions of years: plutonium-239 (half-life 24,000 years), technetium-99 (half-life 220,000 years) and iodine-129 (half-life 15.7 million years). Madres says, "Without a permanently safe location for these byproducts, society will have to carry the burden of storing and guarding nuclear wastes for many centuries. This turns the nuclear energy process into a moral issue involving sustainability and the fact that the power consumed today will leave radioactive garbage for future generations."[108]

Another concern is the vulnerability of storage facilities to terrorist attacks and leakage from geologic repositories that are designed to isolate high-level waste from the natural environment.[109] The problem with nuclear energy and resources is that they generate high-level waste, which has the capability of harming living organisms and the environment over long periods of time. Presently, there are no permanent radioactive waste storage facilities in the United States, and the intermediate-level radioactive waste facilities are where wastes are temporally being stored. Approximately 270,000 metric tons of high-level radioactive wastes have reportedly accumulated in thirty countries, with thousands of additional metric tons added annually. Even the strongest promoters of nuclear energy have not been able to come up with a viable solution to the ultimate disposal place, method, guardianship and long-term management of nuclear waste.[110]

In Who's Backyard?

Nuclear waste must be dealt with in a manner that doesn't affect the public, but where should it go? Waste has to "go somewhere," but where? In whose backyard? And at what cost to neighboring communities and future beings? On October 11, 2017, a press release, "Citizens Oppose HR 3053—Nuclear Waste Policy Act Amendments of 2017," stated that, should Congress pass HR 3053, the Nuclear Regulatory Commission (NRC) would be allowed to license interim storage facilities for the nation's deadliest high-level radioactive waste coming from the Dallas, Bexar, Midland and Nueces County Commissions in Texas and the cities of San Antonio and Lake Arthur, New Mexico. The license for interim storage facilities would allow for "an unprecedented mass movement of high-level radioactive waste across the nation…with over 10,000 train shipments, occurring over a period of 24 years. A previous DOE study found that at least one train accident could be expected if transport was done mainly by that mode. A 42-square mile area could be contaminated from a small radioactive release, and remediation costs could range from $620 million in a rural area up to $9.5 billion for a single square mile in a major city."[111]

Karen Hadden, executive director of Sustainable Energy and Economic Development (SEED) Coalition, expressed her concerns: "Homeowners' insurance policies generally don't cover nuclear accidents and no one wants their children accidentally exposed to radiation that can cause cancer, genetic damage or death. We strongly oppose the John Shimkus Yucca Mountain Nuclear Waste bill. At minimum, it should be amended to require that transportation routes be designated before any consolidated interim storage site for deadly radioactive waste can be licensed. People have a right to know if they're at risk for shipments of radioactive waste coming through their neighborhoods."

Tom "Smitty" Smith, senior advocate for Public Citizen's Texas office, issued this statement:

> *Moving this deadly waste simply to store it all in one place creates unnecessary risks from accidents, leaks and terrorism. There's no need to move it right now and it's less dangerous to secure it at reactor sites. Radioactive waste should only be moved once, when a permanent repository with the right location and robust storage systems is available, in order to isolate the waste for millions of years. Storing dangerous radioactive waste for decades in a seismically active region can only lead to disaster. Millions of people could be impacted by leaks or an accident if the nation's largest aquifer, the Ogallala, became contaminated.*

Elizabeth Padilla, a mother who resides in Andrews County, Texas, also expressed concerns: "We received none of the electricity produced by nuclear reactors in California, Chicago or New York, and had no economic benefit from them. It is a massive injustice to bring in the most deadly radioactive waste from the entire country and dump it in our backyard. They generated the waste. They should take responsibility for storing it where it was generated, and not dump it on communities that don't have the millions of dollars needed to fight back."

Don Hancock, nuclear waste program director at Southwest Research and Information Center in Albuquerque, New Mexico, reflected the history of waste containment facilities in his statement, "For decades, the large majority of New Mexicans have repeatedly said NO to commercial spent fuel when it was proposed by the Department of Energy for the Waste Isolation Pilot Plant (WIPP) and when utilities proposed a Holtec-type dump on Mescalero Apache land. Those proposals were defeated."

Leona Morgan of Diné No Nukes offered a cultural perspective on the issue: "The Navajo Nation prohibits transport of radioactive materials through Diné Bikeyah—our indigenous lands—but doesn't have jurisdiction over federal and state highways or railways. In addition to the thousands of abandoned uranium mines in the Southwest, having high-level radioactive waste transported through and dumped here would add to the radioactive risks people suffer already."[112]

PART II

WATER JUSTICE

5
THE LAST WILD RIVER

Somehow the watercourse is to dry country what the face is to human beauty. Mutilate it and the whole is gone.
—Aldo Leopold, 1937

The Gila River, located in southwestern New Mexico along the Arizona border, is one of the few remaining "wild" rivers in the United States. Human occupation and land use dates back thousands of years, with groups of early inhabitants finding shelter within canyon caves of the Gila basin. The Hohokam civilization built large structures and canals to irrigate fields along the Middle Gila River and Salt River between AD 600 and AD 1450. In the late 1200s, people of the Mogollon culture inhabited the upper Gila basin, creating their civilization and lodging, like those at Gila Cliff Dwellings National Monument in Silver City, New Mexico.[113]

Before the arrival of Spanish explorers, the Gila River was central to a band of Pima, the Keli Akimel O'odham (Gila River People), who lived on the banks of the Gila River and considered it holy. The river has been called by many names, such as Akee-mull, Apache de Gila, Brazo de Miraflores, Cina'ahuwipi (Chemehuevi or Southern Paiute language), Hela River, Jila River, Rio Azul, Rio Gila, Rio de las Balsas, Rio del Nombre Jesus, Rio del los Apostoles, Zila River, Xila River and Keli Akimel. The word *Gila* is thought to have been derived from a Spanish contraction of the Yuma word *Hah-quah-sa-eel* ("running water which is salty")[114]

Archival records indicate Spanish explorer and missionary Juan de la Asunción encountered the Gila River in 1538 while traveling north along either the San Pedro or Santa Cruz tributaries.[115] Hernando de Alarcón, most noted for his expedition to Baja California, traversed the Colorado and Gila Rivers, drawing a map showing the Gila River, which he called the Miraflores or Brazos de la Miraflores.[116]

Leader of the last stand of Apache resistance during the Indian Wars, Geronimo of the Bedonkohe (Be-don-ko-he) band of the Apache was born near Turkey Creek, a tributary of the Gila River near the present-day town of Cliff, New Mexico. According to an interview quoted by David Roberts in the book *Once They Moved Like the Wind: Cochise, Geronimo, and the Apache Wars*, Geronimo remembered his childhood growing up along the Gila: "Sometimes we played hide-and-seek among the rocks and pines; sometimes we loitered in the shade of the cottonwood trees or sought the shaddock [a kind of wild cherry] while our parents worked in the field."[117]

The Gila River served as a part of the border between the United States and Mexico after the Treaty of Guadalupe-Hidalgo of 1848. Then, after the Gadsden Purchase of 1853, the U.S. border was extended south of the Gila to its present borders. By 1871, Mormon settlers began to populate the Gila River valley around present-day Phoenix, using the Gila, Salt and San Pedro Rivers for irrigation for six major settlements.[118]

In the early 1900s, major dams within the Colorado River and Gila River basins were built and operated by the U.S. Bureau of Reclamation. Only one dam is located on the Gila River, the Coolidge Dam, thirty-one miles (50 km) southeast of Globe, Arizona. It holds water at San Carlos Lake.[119] Historically, the Gila River and its main tributary, the Salt River, were perennial streams carrying large volumes of water, but irrigation and municipal water diversions have dried the rivers. The natural discharge of the river was roughly 1,300,000 acre-feet (1.6 km^3) per year, with a mean flow of about 1,800 cubic feet per second (51 m^3/s) at the mouth. Boating and fishing are popular recreational activities at San Carlos Lake and other basin reservoirs; the Gila River is navigable with Class I to III whitewater, during spring snowmelt and after summer and autumn storm events. Many tourists hike through the canyon trails and scenic waterways; much of the region remains roadless.[120]

The greater Gila bioregion is considered an ecological gem and the birthplace of America's wilderness movement. The Gila National Forest was created on March 2, 1899, in an effort to protect the watershed and Gila River from adjacent land use. In 1921, ecologist Aldo Leopold, writing in the *Journal of Forestry*, noted, "Selecting such an area as the headwaters of the Gila River on the Gila National Forest to be conserved as wilderness—a place where industrialized logging, roads and buildings would be barred." Leopold's campaign led to the Gila Wilderness being the nation's first designated wilderness in 1924 and also instigated a wilderness-conservation movement that continues today.

RIPARIAN ECOLOGY

*In New Mexico, water is precious and riparian systems provide the sustenance for life and harboring rich biodiversity. Within the Gila wilderness, endemic species such as the Gila Trout (*Oncorhynchus gilae*), Mexican grey wolves, jaguars, Gila chub, Gila top minow, loach minnow, southwest willow flycatcher, Chiricahua leopard, two species of bat, and the Mexican spotted owl—are all threatened or endangered species. Black bears, cougars, elk, deer, bighorn sheep, turkey, javalina, porcupine, over 30 species of fish, various cacti, grasses shrubs, and trees also populate the Greater Gila Bioregion. Floods recharge groundwater with fluctuating levels depending on the amount of precipitation and duration of storm events and river flow. Large floods tend to scour riverbed channels, moving sediment, creating pools and natural wetlands, and also saturating the floodplain, allowing the growth of dense vegetation, which is critical habitat for bird species such as the Southwest willow flycatcher and yellow-billed cuckoo. The southwestern willow flycatcher, for example, needs moist or saturated soils during the summer months to sustain conditions necessary for successful reproduction—specifically thermoregulation of their eggs and nestlings. Migrating birds that land and breed in the Cliff-Gila Valley are primarily foraging on invertebrate prey that require slow recession rates of surface water and groundwater in the spring, as well as availability of surface and ground water in the hot, dry summer months. Wetlands provide habitat and food for northern Mexican gartersnakes, Sonoran mud turtles, Great Plains toad and where the Woodhouse's toad breed and their tadpoles grow and develop into terrestrial juveniles. Most mammal species need riparian habitats for 95% of their survival needs and therefore rely on herbaceous vegetation as well as wetland conditions. Mid-size flows during winter and the snowmelt-runoff period produces water flows that sort gravel and cobble stones used by native fish throughout their life cycle. As flows recede*

> *during spring, fine sediments are removed from gravel within the stream bed channel, which allows spike dace and loach minnow a habitat in which to spawn. Cobble bars provide habitat for tiny aquatic insects that fish feed on. In an environment with limited water supply, nonnative fish, such as catfish and small-mouth bass, compete for food and prey on native species until monsoon rains restore flows to the river and all organisms benefit from increased food sources and flow.*[121]

Climate change threatens the biodiversity and ecological integrity of the river as well as its forested regions. Off-road recreational vehicles pose a danger to the wildlife habitat of the Mexican gray wolf recovery program, which faces pressure from the livestock industry, which blames predators for livestock loss and promotes contests of killing coyotes and prairie dogs. The organization Wild Earth Guardians believes that the Gila bioregion, which includes 2.2. million acres of roadless land that contains volcanic cones, cliff formations, steep mountain valleys and shallow desert streams should be preserved as "wilderness."[122]

For over two decades, environmental advocates protected the river from construction of two proposed dams. In 2004, the Gila was named one of the nation's most endangered rivers due to the proposed Hooker Dam. New Mexico governor Bill Richardson advocated and blocked any project comprising the portion of the Gila within the state.[123] In 2011, threat to the river was from a proposed water diversion that would have taken water during flood events and diverted an estimated fourteen thousand acre-feet of water per year, piping it some twenty-five miles to an off-channel reservoir. Water would have been taken from an already shallow low-flowing stream, decreasing hydrologic connections to the floodplains that support critical habitat for wildlife and rich biodiversity in the region.[124]

In an article on the Gila River in *National Geographic*, Martha S. Cooper, a forest ecologist and southwest New Mexico field representative with The Nature Conservancy, said, "On the New Mexico side, the river is actually healthier than it was two decades ago. In addition to fighting off the dams, conservationists have worked with local ranchers to reduce the damage caused by cattle grazing along the banks and in the floodplain. So much riparian restoration is about planting trees, but here the river is healing itself. The Gila still has its hydrograph. The floods still have the power to act creatively on the landscape."[125]

The Gila presently suffers from diversions within the downstream reaches below the mountainous Gila Wilderness from valley farmers who leave the river dry for many weeks during the summer to irrigate local farms and ranches. As outlined in The Nature Conservancy's detailed report on the Gila:

> *The river is unique in the variability of its natural flow: The river goes through radical changes in depth and size over the course of a year. This variation in flow is an essential component to a locally flourishing biodiversity. The highly fertile floodplain that surrounds it is dependent on regular inundations of nutrient-rich sediment, the replenishment of groundwater sources and the relative absence of dangerous flooding. These highly mutable seasonal flows support a rich array of plants, animals, invertebrates and fish. Two endangered fish species live in the river: spikedace and loach minnow, and any storage or diversion would not only threaten the last remaining intact native fish populations in the Colorado River Basin, but also reduce the numbers of insects crucial to the health of the entire ecosystem while compromising the availability of water for the cottonwoods and willows.*[126]

According to the 2004 Arizona Water Settlements Act, New Mexico would be required to pay an exchange fee of $122 per acre-foot to divert water. The cost of construction operation fees and purchase of water from the Central Arizona Project (CAP), a federally subsidized diversion of the Colorado River, would have been costly for Arizona cities, farms and the Gila River Indian Community in Arizona. The water diversion project would have cost an estimated $300 million to construct and another $5 million or more per year to operate.[127]

In an article on Truthout.org, "The Battle to Save New Mexico's Last Wild River," Grant County planning director Anthony Gutierrez disputed the mayor of Silver City, Mike Morones's, point about the area having enough water. "We have well-drillers in the area that I speak to pretty consistently, that say wells are going dry all over the place," Anthony Gutierrez said. "So we're losing a substantial amount of groundwater." He noted: "We're getting absolutely no recharge. It hasn't snowed here in probably seven to 10 years. I mean, it snows, but it doesn't get a substantial snowpack. So now we're seeing all of those groundwaters being depleted, especially in the Luna County area, where they have substantial, and a lot more agriculture in the area, huge chili farmers, cotton, alfalfa, and corn." Gutierrez said he was contemplating and planning not for the next 10 years, but the next 50. He projects increases in both population and water use; "without snow, one cannot have water in the West."[128]

The Gila remains one of the few "untamed" true wilderness areas in the American Southwest. An ecological rare beauty, the Gila bioregion contains a river that floods with seasonal fluctuations, providing shelter to the rich diversity of plants and organisms, thereby creating a complex and thriving ecosystem. Off-road vehicles, introduction of roads and the uncertainty of water flow due to climate change threaten the future of the Gila Wilderness. Without this biodiverse wilderness, where would endangered species such as the southwest willow flycatcher, the Chiricahua leopard or the Mexican spotted owl seek refuge? Who will speak on behalf of the native trees and the endemic species of this special bioregion to come? It would be an environmental injustice to lose protection and preservation of the Gila bioregion and America's last "wild" river.

6
RIVER OF LOST SOULS

El miedo de las cosas que no son nos hacen padecer,
y las cosas que son—no paracen ser.
(Fear of the things that are not make us suffer,
and the things that are—we don't believe.)

The San Juan River basin is an area of 9,744 square miles that encompasses portions of Rio Arriba, McKinley and Sandoval Counties. Located in northern New Mexico, with approximately 211 miles of perennial streams containing major tributaries, the Animas and La Plata Rivers contribute to the San Juan River, the largest tributary of the Colorado River. New Mexico receives an average of twelve inches of rainfall per year, with an estimated six inches falling in the New Mexico San Juan River basin area. High elevations and extreme temperatures cause increased rates of evaporation, and spring snowmelt is the primary source of water flowing in the rivers of the San Juan Mountains.[129]

Annually, during winter, the Animas River serves as a feeding ground on ice-free rivers and is the habitat for resident birds and migratory bald eagles.[130] The Animas River attracts white-water rafting, which accounts for 8.9 percent of Colorado's commercial rafting market, while generating $5,207,033 annually.[131] Aquatic insects such as diptera and mayflies occur in the winter, spring and summer months. All through the fall season, caddisfly benthic macroinvertebrates provide food for the population of fishes such as rainbow, brown, Colorado River cutthroat and brook trout that normally thrive in cold mountain streams and rivers. The Animas River is utilized year-round for recreational fishing due to moderate winter weather.[132]

Ancient puebloan irrigation canals and architectural ruins indicate a long history of settlement in the San Juan River basin and the Animas and La Plata River valleys. The most notable archaeological site is the Aztec Ruins National Monument, which is situated along the river in the present-day town of Aztec, New Mexico. The basin is also the homelands of the Ute and Navajos. In the 1500s, Spanish conquerors settled throughout New Mexico, establishing the acequia system for control of water and irrigation. When New Mexico became a territory of the United States in 1848, the territorial legislature began establishing water law in New Mexico based on the Indian and Spanish practices until 1912, when the Office of the State Engineer began facilitating and administrating water rights.[133]

LEGEND

One legend of how the Las Animas River got its name dates to the early 1500s, when Spanish conquistador Francisco Vázquez de Coronado y Luján led an exploratory expedition into northern New Mexico and a group of his men died along the banks of the river. According to Catholic belief at the time, the souls of individuals who died without last rites would go to purgatory. Without a priest or church to conduct last rights for Coronado's men, it was believed that their souls were lost or at a perpetual state of unrest, haunting the river, thus the river was named El Río de las Ánimas Perdidas en Purgatorio (The River of the Lost Souls in Purgatory). Another legend says that early Spanish explorers noticed large numbers of ancestral pueblo ruins by the river and felt haunted by their emptiness as though forgotten and named the river Rio de las Animas Perdidas.[134]

The San Juan basin, located on the Colorado Plateau, possesses few roads within the rugged mountains, most of the region being above five thousand feet in elevation. As the region grew in population and development, the Upper Colorado River Basin Compact was created in 1948, and the Bureau of Reclamation constructed the Navajo Unit of the Colorado River Storage Project (CRSP). Under this compact, the Navajo Dam was completed in 1962, constructed on the San Juan River with 1,708,600 acre-foot storage capacity reservoir that extends into the state of Colorado. The Navajo Dam provides water to the Navajo Indian Irrigation Project, which is said to irrigate 110,630 acres of alfalfa, corn, wheat, barley, potatoes, onions, pinto beans and pasture lands on the Navajo reservation.[135]

The Animas River contributes water flow to the San Juan River, which supplements the flow in the Rio Grande River basin by diverting an average of 110,000 acre-feet of water annually. This San Juan–Chama Project is made possible by a series of tunnels under the Continental Divide. The water diverted is used for irrigation in the Rio Grande basin and for municipal and industrial uses in Santa Fe, Albuquerque and adjoining cities. In 2015, the Animas–La Plata Water Project was completed and now pumps water to fill the Lake Nighthorse reservoir in compliance with Southern Ute tribal water rights claims associated with the Colorado Ute Settlement Act amendments of 2000. Water from the San Juan, Animas and La Plata Rivers serve an estimated population of 128,529 in the cities of Aztec, Bloomfield and Farmington and the unincorporated rural areas of San Juan County.[136]

HISTORY OF CONTAMINATION

The Animas River has a long history of contamination starting in the 1870s, when Anglo settlers began mining for gold and silver. Before the Gold King mine spilled three million gallons of toxic wastewater into the river, mine tailings had been dumped and the river tainted. Toxic water had been discharged into the Cement Creek at a rate of 50 to 250 gallons per minute, or more than 100 million gallons per year. The Animas River picks up further contamination as it flows through the nation's most prolific natural gas fields and coal bed methane wells, then through the San Juan area between two mega coal-fired power plants, and onto areas where uranium is mined as well as the Aneth oil field. Farming, grazing and urban communities are also the cause of high nutrient loading and elevated bacteria levels in the Animas and San Juan Rivers.[137]

> *In 1965, a tailings pile located on the banks of the Animas River northeast of Silverton was found to be spilling cyanide-laced water into the river. In June 1975, the same tailings pile was breached which dumped tens of thousands of gallons of waste tailings water, along with 50,000 tons of heavy-metal-loaded tailings into the Animas for 100 miles downstream. According to a* Durango Herald *reporter at the time of the spill, the river "looked like aluminum paint" and fish that were placed in a cage into the*

water near Durango all died within 24 hours. In 1978, mine workings got too close to the floor of Lake Emma, lake water burst through sending an estimated 500 million gallons of water through the mines, sweeping up huge machinery, tailings and sludge, and blasting it out the American Tunnel sending everything downstream. By the 1980s, mining waned and in 1991 the last major mine in the San Juan Basin was shut down, and there were an estimated 400 mines within the watershed with many of them having unmitigated discharges into streams. The impact on the aquatic environment was evident with no fish present in streams.[138]

The Gold King mine was considered for Superfund status, but some of the local merchants feared that the stigma of this designation would destroy tourism and end the possibility of future mining within the area. Concerned citizens and representatives from industry and federal and state agencies created the Animas River Stakeholders Group in 1994 and began to address concerns regarding water quality, property values and economic impacts of the river's contamination on industry. When the mine closed in 1991, Sunnyside Gold Corporation planned to plug, or bulkhead, the mine, which would be a short-term fix. In the long-term scenario, water and toxic contaminants were likely to accumulate and find another pathway into the environment. The state refused Sunnyside's plug plan, and the company agreed to clean up abandoned mines nearby and run the polluted waters through a water treatment facility. By the early 2000s, water quality levels showed improvement and metals detected in the river had been reduced by 50 to 75 percent.[139]

Sunnyside turned over its water treatment operations to Gold King, which hoped to resume mining operations in the area. After Sunnyside sealed its bulkheads at the Red and Bonita mines, water backed up and started pouring out of natural fractures. The operation of mining waters being run through the treatment plant, which Sunnyside left behind, ceased when Gold King ran into technical, financial and legal troubles and the treatment plant could not continue operations. Environmental deterioration ensued and fish that had

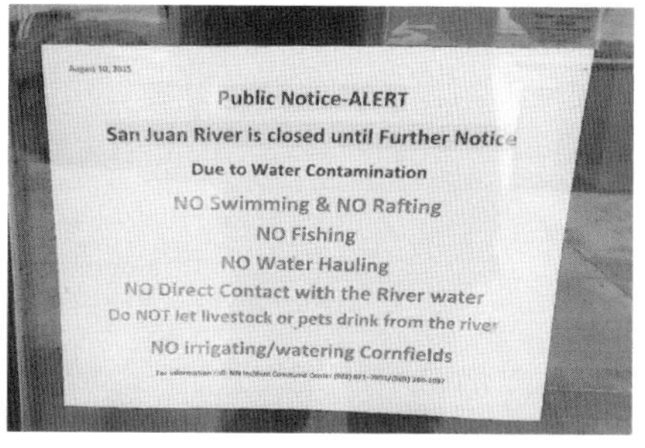

Navajo Nation storefront window public notice regarding the San Juan River after the Gold King Mine spill. *Photo Credit: Earl Tulley.*

returned to the Animas River below Silverton were wiped out once again. If the site had been designated a "Superfund," the water treatment plant would have resumed operations until a plan was created to deal with long term contamination from the closed mine within the basin. Until the Gold King mine is permanently plugged, acid mine drainage will continue to accumulate with inevitable spills and leakage to ensue.[140]

THE TOXIC SPILL

According to a *High Country News* article, "Animas River Spill: Only the Latest in 150 Years of Pollution. Mapping the Other Threats to the Animas and San Juan Rivers":

> *On Aug. 5, 2015, approximately three million gallons of wastewater and toxic sludge poured out of the dormant Gold King mine into Cement Creek, a tributary of the Animas River. The waters turned milky green, then yellow, and finally a thick cloudy orange or mustard color. Water had backed up in the Gold King mine behind a dam which formed when the mine portal's ceiling collapsed at some point earlier in time. The Environmental Protection Agency had assessed the mine's condition and had planned to install a pipe to drain the water so that they could eventually plug the mine, keeping the contaminated water inside it and out of the streams. Unfortunately, there was a breach in the dam and the contaminated water was released.*[141]

Dave Tomko of the San Juan Watershed Group hypothesized that naturally high pH levels of the San Juan Basin's geology would buffer acidic toxic releases. But his major concern was that crops would dry up when irrigation was shut off shortly after the Gold King mine spill in August 2015. The city of Durango, which uses the Animas River to provide for a portion of its drinking and irrigation water, was most economically impacted by the spill, as was its tourism. Recreational activities slowly diminished as officials closed the river for public health reasons, shutting down hundreds of recreational boaters, tubers and rafting industries.[142]

Environmental Impacts

The fate of the fish, birds, bugs and other wildlife that call the river home is uncertain. When water comes into contact with iron disulfide (pyrite) and oxygen is present, sulfuric acid forms, creating acidic water, which dissolves naturally occurring heavy metals such as zinc, lead, cadmium, copper and aluminum, resulting in contaminated water that flows out of mined areas. Acid mine drainage traverses fractures, cracks and tunnels within the mountains containing mines and drains downstream into natural hydrological pathways, eventually joining tributaries that contribute to larger bodies of water or settle into lakes.

Color is an important indicator of water quality. It is psychologically distressing to see an orange or mustard hue to a natural flowing river whose water flow stems from snowmelt. Water in some domestic wells near the Animas River reportedly had a yellow tint in the days after the spill. Irrigators had to shut down their ditches during the hottest months of the year, causing crop damage. The toxic spill flowed through critical habitat for sensitive species such as the razorback suckers and pike minnows.[143]

Concerns over the effects of the Gold King mine spill and the health of neighboring communities is of consequence. Toxic metals from mining include "heavy metals," which are individual metals such as arsenic, selenium, chromium, mercury and lead and/or metal compounds that can negatively affect human health. Small amounts of heavy metals form the building blocks that support life; however, large doses have proven to be toxic. High levels of exposure to arsenic can cause death; beryllium can cause sensitization as well as lung and skin disease in a significant percentage of those exposed; and calcium chromate, chromium trioxide, lead chromate, strontium chromate and zinc chromate are known human carcinogens. Mercury and lead exposure are also detrimental to human health and can lead to permanent nerve and kidney damage.[144]

Cleanup

EPA officials admit that before tinkering with the mine, they felt that they should have taken better steps to mitigate a possible disaster, such as drilling into the mine from the top to assess the situation without danger of busting the dam. It is likely that water that had accumulated in the mine might have eventually escaped through natural fractures of the rocks or built up pressure over time then busted through the dam at any given moment. In 2015, the EPA considered declaring the area around the Gold King mine a Superfund site and has taken responsibility for the accident, spending $29 million to initiate cleanup.[145]

Prior to the spill, the stream was healthy enough for agriculture and other uses. As a result of the spill, "irrigation lines were cut off. Corn, melons, hay and wheat never made it to market. The spill delivered a psychological lashing in a drought-stricken place where water is gold, many live in poverty and the San Juan River holds financial and spiritual power."

The Navajo Nation filed a lawsuit alleging mismanagement of the mine that resulted in 880,000 pounds of metals spilling out when the Gold King burst. The lawsuit names several defendants: the EPA; two contractors (Environmental Restoration and Harrison Western); four mining companies (Gold King Mines Corporation, Sunnyside Gold Corporation, Kinross in Canada and Kinross USA); and ten unnamed individuals. The suit also claims that "roughly 80 to 90 percent" of the toxic waste spills remain embedded in the river upstream, ready to spill into the Navajo Nation during rains and storm flooding events. Near the Gold King mine, toxic water continues to flow out of the mine at a rate of 570 gallons a minute.[146]

In December 2016, the Navajo Nation filed a claim with the federal government seeking $162 million in costs from the Gold King mine spill, which included costs such as $3.1 million for unreimbursed expenses and $159 million to develop alternative water supplies, future monitoring and other costs. Although the San Juan River was adversely affected by the Gold King mine spill, the EPA rejected the Navajo Nation's costs of more than $250,000 to haul drinking water, which could not be supplied by the San Juan River. The EPA agreed to pay $4.5 million to state, local and tribal governments for their emergency response following the spill but turned down $20.4 million in other requests for past and future expenses. The agency did agree to pay more than $90,000 to transport water to two areas until early September 2015, when the river was said to have returned to pre-spill water quality conditions.[147]

The State of New Mexico sued the EPA and the State of Colorado over the accident; however, on June 26, 2017, the *Santa Fe New Mexican* newspaper reported that "the U.S. Supreme Court rejected the lawsuit that claimed that the EPA played a direct role in the spill." Downstream communities of the Gold King mine and other spill sites are feeling an economic impact due to the loss of revenue from polluted agricultural lands. Outdoor recreational activities have decreased after the Gold King mine spill of 2015, and the expenses from contaminated wells, river water and hauling water for domestic purposes has further burdened the communities with unforeseen costs.

"The River of Lost Souls" will continue to be haunted, perhaps not by the Spanish lives that were lost centuries ago, but by the legacy of mining and other human land use activities that have marred the once-pristine river. The river itself continues to be threatened with further risk of environmental degradation from abandoned mines. While the Gold King mine spill was deemed "accidental," better planning and management of the volatile mine by the EPA, could have reduced the environmental impact. Pollution cleanup was never considered a priority in this region by the EPA, and mine owners and/or the companies that owned the mines were not held responsible for cleanup practices following industrial use.

For the past 150 years, mining operations have dumped waste into the Animas River, wreaking havoc on the wildlife, aquatic organisms, the unsuspecting eagle migrants and all who reside along the river. This "sparsely populated" area contains Hispanic, Ute, Pueblo of Picuris, Navajo and Apache impoverished communities of color who consider the water and mountains to be sacred. Until recently, the Navajo and Ute Nations have not had the political clout to oppose the U.S. government's permits to mining or file lawsuits regarding contamination in the San Juan basin watershed.

On Monday, February 12, 2018, the Navajo Nation won a battle for fair compensation claims. "The United States District Court for the District of New Mexico denied contractor Environmental Restoration, LLC's motion to dismiss the Nation's claims, instead upholding all claims, including CERCLA claims, and claims for negligence, gross negligence, trespass, and nuisance. The Court also refused Environmental Restoration's demand to strike the Nation's request for punitive damages."[148] In response to this announcement, Navajo Nation president Russell Begaye stated, "We will continue to fight for justice for the Navajo people and the Navajo Nation. Our people have suffered greatly and must be compensated fairly." Navajo Nation attorney general Ethel Branch also remarked, "This is an important landmark in our fight to hold the parties responsible for the harms caused by their negligent and reckless conduct. We will continue to push ahead with renewed strength and resolve."

7
URBANICIDE OF BURQUE

The Villa de Albuquerque was founded by Spanish colonists in 1706 in what is present-day "Old Town," approximately two miles west of the railroad. As the city expanded with the arrival of the railroad in 1880, the original Villa de Albuquerque along with the Hispanic villages of Barelas, Martineztown and Santa Barbara were incorporated to form a larger metropolis. Downtown Albuquerque became a thriving community and a hub at which goods were bought and commodities sold. People sought health care and worked downtown, with agricultural fields located along the outskirts of town. Albuquerque, historically a trade stop along the Camino Real, was an economically viable station for the AT&SF. Large locomotive maintenance shops and regional administrative offices were located in the heart of the Barelas community. The railroad became the area's largest employer of physical laborers in the heart of Albuquerque.[149]

For centuries, the Middle Rio Grande valley irrigated fields using acequia systems. New technology and machinery such as steel plows and mechanical reapers harnessed more production and product yield for valley farmers, growing farms in size and capacity. Sheep farming and wool production expanded, and designated grazing lands pushed into the Estancia Valley and to the west along the Rio Puerco.[150]

The cattle industry also grew; large herds were kept in confined areas with protective corrals carefully delineated with barbed-wire fences that wreaked havoc on the sensitive grassland ecosystems of the surrounding mesas. Livestock hides, wool and meat were shipped to eastern markets. Brickyards,

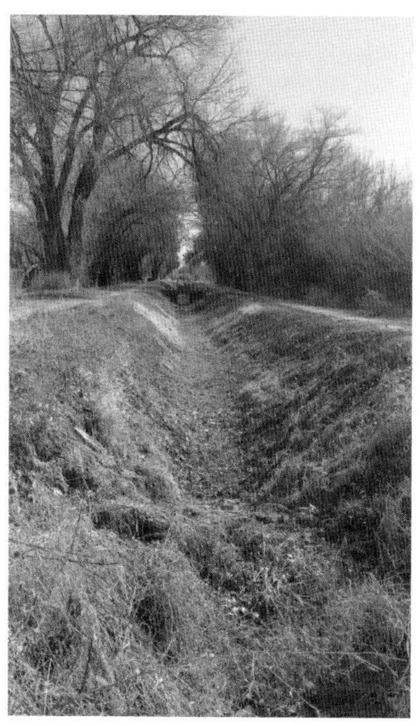

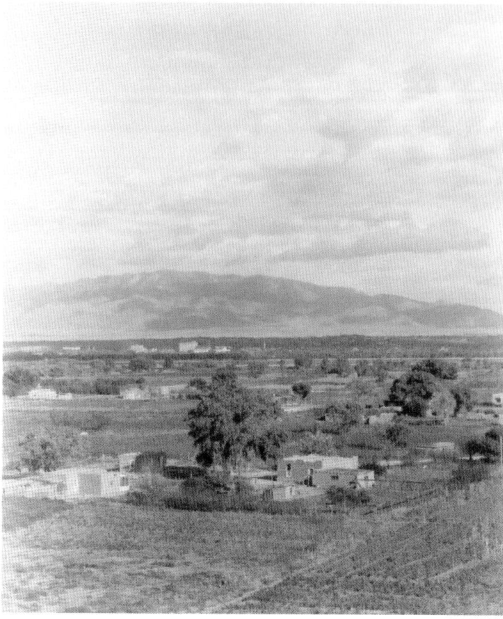

Left: Historic Albuquerque acequia. *Photo credit: Valerie Rangel.*

Right: South Valley farm scene, Albuquerque, New Mexico, circa 1930. *Photo credit: The Albuquerque Museum.*

tanneries, flour mills, packinghouses, wagon factories, steam laundries, bottling works, ice companies, cement plants, wool mills and warehouses were constructed close to the railroad tracks. In 1901, the American Lumber Company purchased timberlands in the Zuni Mountains and built a sawmill along with a woodworking factory. These were connected by railroads. Timber was processed in Albuquerque and then shipped to markets in the West.[151]

ABQ's South Valley Contamination

The Pedro Varela (Barela) family established a farming community in 1662 that lay astride the Camino Real. This point was used widely as a river crossing.[152] The railroad track and locomotive shops that were built in today's Barales neighborhood cut off the Varela traditional farming communities' connection to their eastern agricultural fields and main acequia, effectively ending the agricultural heritage and employments that had been established.

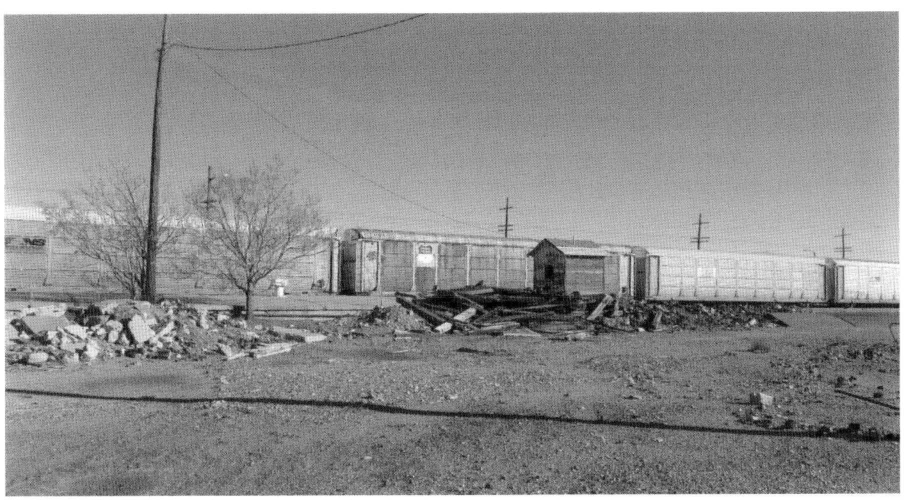

Pile of railroad ties, Albuquerque, New Mexico, February 4, 2017. *Photo credit: Valerie Rangel.*

After the Depression, personal automobiles and truck transportation of material goods became the new focus. Roads were built and the state highway network improved. Rather than a way for mainstream Americans to travel across the country, the railroad became a vehicle to transport military troops and materials from coast to coast. In the 1950s, an industrial boom expanded development in the South Valley of Albuquerque. Without environmental regulations or zoning in place, South Valley residents soon faced issues of contaminated wells and mysterious illnesses.[153]

The history of environmental pollution in the South Valley of Albuquerque dates to the early 1900s. From 1908 to 1972, the Atichson, Topeka and Santa Fe Railway soaked railroad ties in the South Valley. The pressure-treatment plant used creosote, oil mixtures and other chemicals to pressure-treat wood products, including railroad crossties, bridge ties, switch ties and other materials. The plant was closed in 1972 and was designated a Superfund site in 1994. According to the EPA, between 59,000 and 70,000 gallons of toxic non-aqueous material slowly dissolved into the ground during the plant's period of operation. In October 2006, the EPA's planned cleanup included soil removal with incineration off-site, soils with no sludge would be capped, and groundwater pumped, treated and re-injected back into the aquifer. It is important to note that in order to make polluted water drinkable again, expensive processes like reverse osmosis or desalination water treatment are needed.[154]

In the 1950s, American Car and Foundry, under contract with the Atomic Energy Commission, began conducting research and experiments in the South Valley to develop nuclear engines for the government. The site where experiments were conducted is now owned by General Electric. In October 1999, a lawsuit was initiated by Patricia Madrid alleging that the plant permanently ruined groundwater in the region.[155]

The Tijeras Arroyo, which extends from the base of the Manzano and Sandia Mountains, travels through Kirtland Air Force Base and also the Mountain View Neighborhoods, transporting surface water runoff into the floodplain and river of the Rio Grande. This arroyo is thought to be the illegal dumping grounds of the military, with undocumented illnesses reported among its residents. The *Albuquerque Journal* reported that the major polluters of the Mountain View neighborhood included three EPA Superfund sites, thirty-two out of thirty-six polluting industries regulated by the EPA, the Public Service Company, Rek Chemical, a water treatment facility, a dairy, more than twenty-five auto recycling yards, five gravel and concrete companies, a solid waste landfill, a fertilizer facility, a chicken farm and more than sixteen major air-polluting industries, as well as sixty-six smaller polluting industries in the area.[156]

Below the Mountain View neighborhood is a nitrate plume three quarters of a mile wide and thirty feet deep that contaminated sixty-one private wells and two city wells. The Superfund sites within the neighborhood are a petroleum hydrocarbon plume from the Cheveron, Texaco and ATA Pipeline tank farms and the old GE plant site. Other industrial operations have further contaminated the soil and groundwater with chlorinated solvents and other hazardous chemicals. The designation of industrial zoning operates on a different standard of cleanup and permitted types of land use than residential zoning; Mountain View is zoned as both industrial and residential.

In 1963, an article in the *Albuquerque Journal* reported nitrate levels ten times higher than safety standards and warned residents, pregnant women and children under one year old to drink only bottled water. South Valley residents had to dig deeper wells due to previous contamination from the sewage treatment plant as early as 1958.[157]

Cleanup for a designated "Brownfield area" or zoned industrial zone is at a standard lower than for residential zoned areas. In essence, residents must choose to live in an area with a history of environmental contamination, reside next to industrial polluting industries, remain in an area with impaired and contaminated water (even after remediation) or relocate in the hope of finding affordable housing elsewhere. Many generations of traditional farmers continue to live in areas of the South Valley, culturally tied to connections of land through their family history.[158]

URBAN SPRAWL

By 1970, the population of Albuquerque skyrocketed and new "urban sprawl" housing developments had sprung up in the Northeast and Southeast Heights and at the foothills of the West Mesa and Sandia Mountains. Strip malls popped up in these neighborhoods, and large commercial retailers decentralized from downtown areas to form two major shopping centers: Coronado and Winrock Malls. The primary places of employment were businesses located around the University of New Mexico, Kirtland Air Force Base, Sandia National Laboratories and industrial parks located next to the interstate highways. In the 1970s, the city built a large-scale sewage treatment plant in the South Valley; residents were forced to tolerate noxious odors and a new source of pollutants in their well water.[159]

The Santolina Master Plan proposed development on the West Mesa of Albuquerque promises to create a 38,000-unit housing development project that is expected to house 95,000 people and provide an estimated 75,000 jobs, complete with open spaces, schools and an industrial area. In a hearing at the County Planning Commission on May 28, 2014, a representative from Consensus Planning declared, "Albuquerque Water Utility Authority has indicated that there is ample water supply to meet the area's needs over the next forty years." Projected water use levels from the Middle Rio Grande Conservancy District (MRGCD) shows an "Excess Demand in 2030" of 46,602 acre-feet per year and does not include estimated water use from the Santolina Development.[160]

The South Valley Regional Association of Acequias (SVRAA) is concerned that development on the West Mesa would seek to acquire and "retire" existing water rights from the *parciantes* of irrigated lands in the Middle Rio Grande Administrative Area. With increasing drought conditions and global warming trends, the City of Albuquerque is unsure of its future water supply and of the future water supply of the Santolina development and its anticipated population.[161]

Cultural Resources

The Petroglyph National Monument is a seventeen-mile-long volcanic basalt escarpment encompassing approximately 7,236 acres (29.28 km^2) located on the West Mesa of Albuquerque. It has a long history of occupation. Indigenous people of the region offered prayers, gathered medicinal plants, used the basalt rocky terrain as a stronghold defense site and considered the landforms sacred. Of the monument's twenty-five thousand petrogylph images, an estimated 90 percent were crafted by Rio Grande Pueblo people who lived and farmed the area from about 1300 through the late 1680s, with the remaining from early Spanish settlers.

In 1997, the U.S. Senate returned 8.5 acres of federal land to the City of Albuquerque for construction of the Paseo del Norte road extension; never before had a national monument been split by a major roadway. Archaeologists confirmed that the proposed road construction would damage at least fifty petroglyphs. Pueblo leaders said that the region offers sacred protection and should not be violated. "It is like an umbilical cord for the tribes around here," said Arden Kucate of Zuni Pueblo. "It ties into our songs and our prayers."

When the national monument was established in the early 1990s, it was on the far outskirts of the city. Now, some homeowners have petroglyphs in their backyards. Currently, the monument is cooperatively managed by the

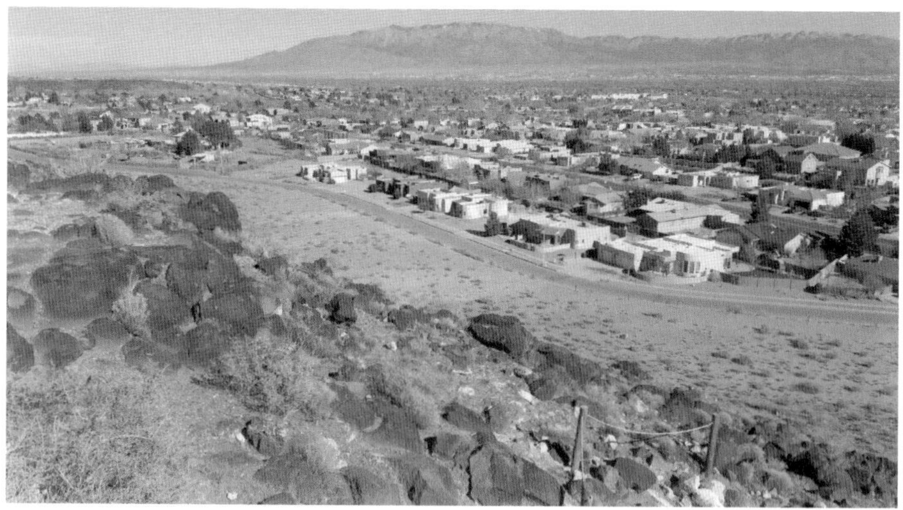

Petroglpyh National Monument at Boca Canyon, Albuquerque, February 4, 2017. *Photo credit: Valerie Rangel.*

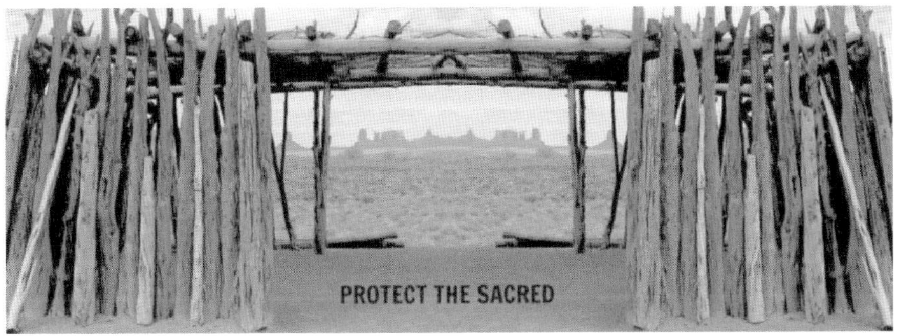

Above: "Protect the Sacred" advertising design on Albuquerque city bus, 2018. *Photo credit and public art by: Nora Naranjo Morse.*

Right: Vandalism at Petroglyph National Monument at Boca Canyon, Albuquerque, February 4, 2017. *Photo credit: Valerie Rangel.*

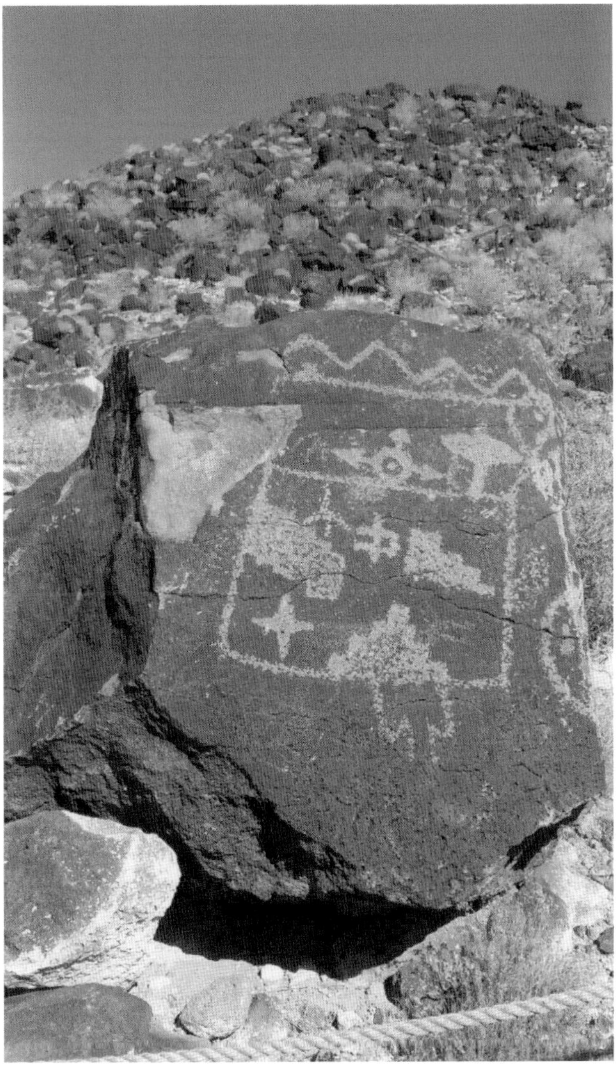

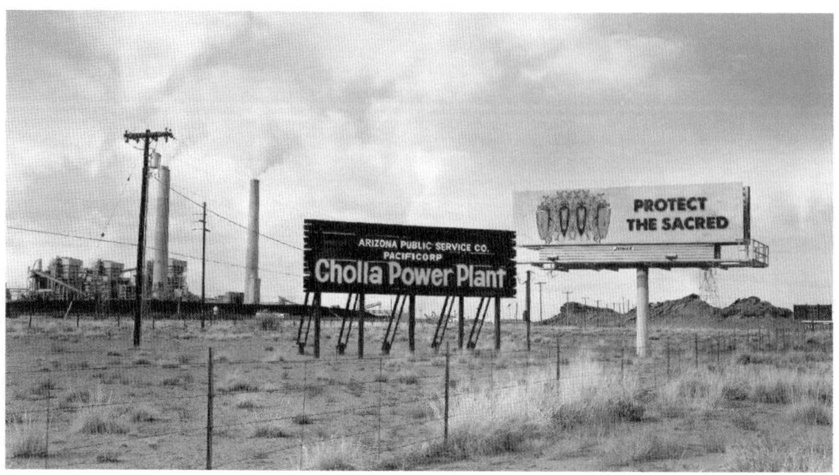

"Protect the Sacred" billboard at Cholla Power Plant. *Photo credit and public art by: Nora Naranjo Morse.*

National Park Service and the City of Albuquerque, allowing recreational use and tourism activities to take place. These activities are opposed by the native peoples of this historical area.

The Petroglyph Monument serves as a reminder of the dynamic history of physical, cultural and psychological connections of the communities within the region. Petroglyph vandalism continues to be a growing problem throughout the Southwest. Rock images of petroglyphs have been chipped away, stolen and sold for profit. Additional traffic and urban sprawl will lead to further vandalism and desecration of this national treasure and impede on the right and access for native peoples to conduct their religious ceremonies.

Nora Naranjo Morse took a photo of Monument Valley and created a design with the text "Protect the Sacred." It will travel as advertising on the outside of two Albuquerque city busses during the spring months of 2018. Morse explains, "Since the beginning of time, humans have been leaving marks on the land. From rock art carved by ancient peoples to interstate billboards advertising cheap gas. We humans have been leaving our mark at different time periods and for a variety of reasons." Nora's billboard situated in front of the Cholla power plant and city bus advertising are public art endeavors with an environmental justice message. She clarifies, "It's the concept of protecting the sacred-ness of life no matter who we are, where we're from or, where we're going."

LEGACY OF TOXIC DUMPING

South of the city, Kirtland Air Force Base (KAFB) contains four landfills containing buried heavy metal and radioactive test animals along with decades of jet fuel leakage that has seeped into the underground aquifer. Heavy nitrate contamination is most likely the result of the wastewater treatment plant, chemical fertilizers, a chicken feed lot and military waste located in the South Valley. In the early 1950s, the base replaced leaking tanks and aging pipelines with a new fuels facility in preparation of jet fuel and aviation gas shipments that started in 1953. An estimated twenty-four million gallons of aviation gas and jet fuel spilled for decades from the pipeline at Kirtland's bulk fuel facility. Ethylene dibromide (EDB) was used as an anti-knock chemical in aviation gasoline and is a known carcinogen. Citizen Action New Mexico, a nonprofit organization based in Albuquerque, led efforts to address the toxic plume of EDB from the spill that has moved in the aquifer to less than one mile from Albuquerque's most productive drinking water wells.[162]

A current estimate suggests that a plume of EDB-contaminated groundwater 1,000 feet wide and more than one mile long is moving northeast from the base; according to a 2009 KAFB memo: "It's advancing as much as 385 feet per year with 80 percent now beyond the boundary of the base and headed towards the Ridgecrest residential neighborhood. The Albuquerque Bernalillo County Water Utility Authority operates a series of wells in the Ridgecrest area that pump so much water for the city that they produce a cone of depression which acts like a straw, sucking the plume ever closer."[163]

From 1980 to 1982, a Texas company illegally dumped PCBs and volatile organic compounds at the western edge of the South Valley in what is now the Walmart site located on South Coors Boulevard. These hazardous chemical compounds infiltrated fourteen feet below the surface in less than two years. Emergency Superfund funds were used to clean up the Pronto site; the company was subsequently indicted and fined.[164]

North of Albuquerque, the Intel Corporation opened a semiconductor fabrication plant in Rio Rancho, supplying high-paying jobs in the 1980s. The corporation picked Rio Rancho as the site for its massive computer chip manufacturing operation and is one of the largest semiconductor companies in the world, with more than four million square feet of manufacturing and office space and employing more than five thousand workers.[165]

In the early 1990s, local residents complained of bad odors emanating from the site and reported symptoms such as chronic headaches and coughing, shortness of breath, nausea, vomiting and even seizures. Other reported illnesses included body rashes, adult-onset asthma, endocrine and reproductive disorders, memory loss, blindness and periodic unconsciousness. Residents claim that illnesses such as these were not experienced until after operations began in 1993.[166]

The federal Occupational Safety and Health Administration classifies crystalline silica as a carcinogen that can scar lung tissue when breathed, causing chronic and acute lung disease that can lead to shortness of breath, weakness, weight loss and death. Intel currently has a permit that allows the manufacturing plant to release 5.9 tons of phosgene each year. Toxic chemicals released by Intel's semiconductor fabrication plant are drawn down into the Corrales Valley and often remain stagnant in the air, trapped by a bluff escarpment. Some speculate that a chemical used in semiconductor factories, hexamethyldisilazane, which is burned by thermal oxidizers then released into the air, is responsible for a fatal lung disease called pulmonary fibrosis that a few residents have been diagnosed with. "It turns into crystalline silica," says Barbara Rockwell, a resident of nearby Placitas who wrote a book about the environmental impact of the Intel plant, *Boiling Frogs: Intel vs. The Village*.[167]

8
DAVID AND GOLIATH

Isleta Water War

The environmental issue is a very spiritual one and I think it takes the understanding and recognition that water and all natural resources are really spirits. Indian people recognize that and so when those spirits are wounded or soiled, if you will, then it's very much a degradation of that particular spirit. And we're all paying for it. So it's very important that we recognize the power of these spirits. We have no control over them. They are very powerful and if we're not real careful how we treat them, we may be seeing some very serious end to us, because the spirits can only put up with so much. They are very real. I think after a long time of being underground with our religion and many of our sacred beliefs, I think Pueblo people are finally beginning to become a little bit more open about it. We are beginning to share it a little bit more, because we're finding that it's very important that the world recognizes the spirituality of it all.[168]

—Verna Olgin Teller (excerpted from a talk given at the 1998 Bioneers Conference in San Francisco)

In the United States, there are 562 federally recognized American Indian and Alaskan native tribes. Isleta Pueblo is one of 19 Pueblo tribes in New Mexico. The village, or pueblo, of Isleta is situated in a valley at the southern base of the Sandia Mountains, with the Rio Grande running down the middle of it. The reservation is approximately 250,000 square acres, located thirteen miles downstream from the City of Albuquerque. Adjacent to the reservation, on the northern boundary, are Sandia National Laboratories and Kirtland Air Force Base.[169]

Isleta is a Spanish word meaning "little island." In 1612, the Spanish mission of San Agustín de la Isleta was built by Catholic Franciscans within an already established pueblo community. During the Pueblo Revolt of 1680, many pueblo people fled to Hopi settlements in Arizona, while others followed the Spanish retreat south to present-day El Paso, Texas. During the Spanish reoccupation of New Mexico, Isleta puebloans returned to the pueblo, during which time members of Laguna Pueblo and Acoma Pueblo joined their community. In the 1800s, the small community of Oraibi was established. As of 2000, Isleta Pueblo included Oraibi and the Chicale community, representing a population of about twelve thousand—four thousand tribal members and an estimated six to eight thousand people who reside on pueblo grounds.

Inspired by an Acoma Pueblo war veteran named Miguel Trujillo, who sued the government and won Native Americans the right to vote, and a growing dissatisfaction with the way tribal government was being run, Verna Olgin Teller decided to become the first female candidate for governor of Isleta Pueblo in 1986. She became the first female governor of a pueblo, serving from 1986 to 1990. Despite some tribal members' efforts to prevent Verna from running and an attempt to overturn the election on grounds that women should not hold office—a traditional belief from the tribe's historical matrilineal clanship system—her election by popular vote withstood.

Teller made radical reforms, including an amendment to the pueblo constitution that led to a change from the tribe's governmental matrilineal clanship system to a tribal council being elected by voters instead of appointed by religious leaders. Under Teller's leadership, Isleta Pueblo became the first sovereign tribal nation to assert its right under federal law to establish its own water quality standards, which meant that the tribe would have authority over upstream discharges from the City of Albuquerque. This was an unprecedented victory in environmental rights. The issue went all the way to the U.S. Supreme Court, where the Pueblo of Isleta prevailed.

As Albuquerque's population skyrocketed, with an increase of 68 percent, from 485,430 residents in 1980 to 712,738 residents in 2000, Isleta Pueblo, being in such proximity, noticed an increase in environmental degradation. Water discharged from the city immediately enters the reservation downstream, where acequia systems divert river water to irrigate and sustain agricultural plots located within the center of the pueblo.[170]

Verna explains: "Our community is a farming community. Our people have made their living for many centuries by farming and so we are looking at the effects of that water coming into the river, into our irrigation canals, into our ditches and eventually watering our food, and eventually getting into the food chain."

James Abeita is a pueblo native and tribal councilman who lives on a small farm just west of the Rio Grande about four miles south of Albuquerque. He's one of many Isleta Pueblo farmers who use water from the river to grow crops like chile, corn and squash. Water pollution is a consistent worry for Isleta farmers, who find themselves trying to reconcile their belief that the river is sacred with their concerns about contaminants washing downstream.

Since the dramatic population growth, religious elders at the pueblo have become increasingly concerned that the Rio Grande water has become too dangerous to consume or to conduct their religious and traditional ceremonies. The tribe conducted water quality sampling and found that the river water contained low-level radioactive waste, which has been discharged into the river from reference labs, hospitals and Sandia Labs. Therefore, any contact with this contaminated water would be dangerous and a risk to the public health of the tribe.

Between 1989 and 1990, Sandia National Laboratories approached the City of Albuquerque Council requesting amendments regarding its wastewater quality standards to allow for more low-level radioactive waste to

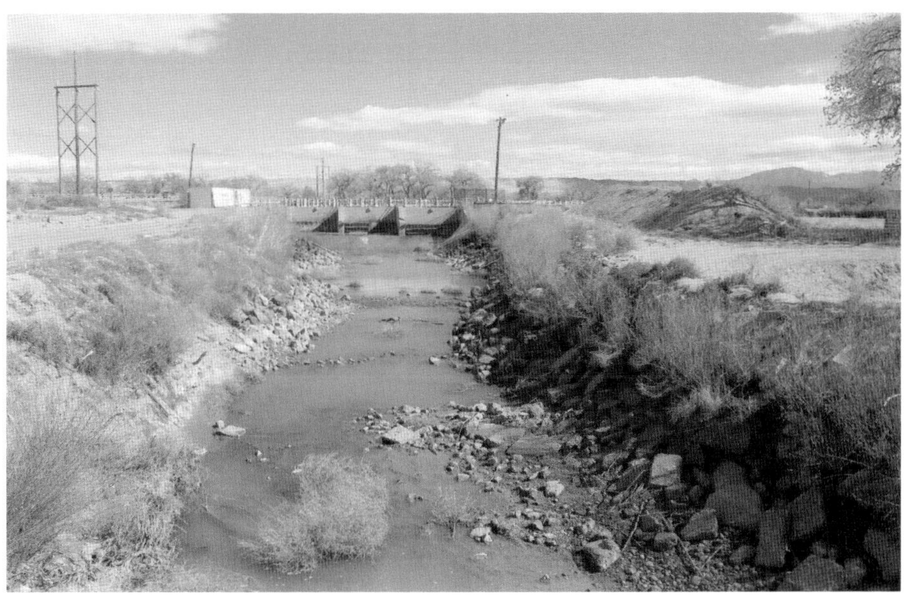

Rio Grande water diverted into an acequia at Isleta Pueblo, February 7, 2017. *Photo credit: Valerie Rangel.*

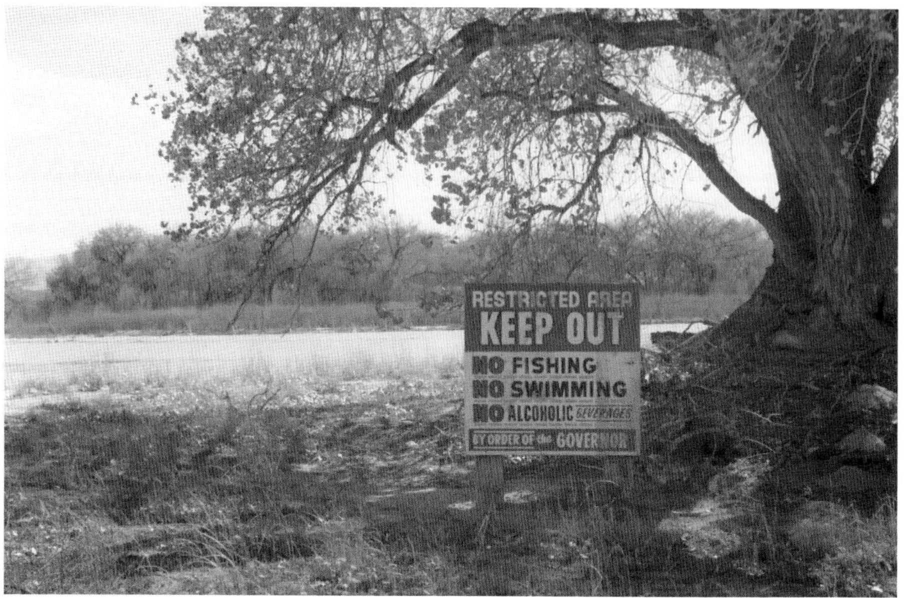

Rio Grande riverbank at Isleta Pueblo, February 7, 2017. *Photo credit: Valerie Rangel.*

enter the Rio Grande. This allowed Sandia National Laboratories to create and experiment with different kinds of weapons using radioactive materials, disposing of contaminated water downstream.

Five miles upstream from the Isleta Pueblo boundary is the City of Albuquerque's discharge point for its water reclamation plant. All wastewater coming out of the waste treatment facility plant in Albuquerque immediately enters the reservation. Albuquerque dumps approximately fifty-five million gallons of wastewater a day and sewage from more than 700,000 residents.

Environmental stewardship is part of the Native American philosophy. Manifesting in every indigenous culture are core values: reverence for nature and reciprocity. Everything has spirit and is alive. The Isleta Pueblo's concern regarding water quality is that it is vital to every aspect of their way of life, from farming to religious ceremonies.

Verna Teller stated:

> *At the end of each summer we complete a month-long ceremony in Isleta Pueblo, which is a combination of the solstice ceremony, leading us into fall and winter, and a harvest ceremony. This particular ceremony lasts for a month and culminates with dancing, cleansing and bathing in the river and*

ingesting river water. We have a number of other very sacred ceremonies that also require ingestion of that water or to immerse in it. Our people have made their living for many centuries by farming. Water continues to be diverted from the river, into irrigation canals, and ditches watering centuries of farm fields that supply food, to the Pueblo.

WATER QUALITY STANDARDS

In 1987, Congress amended the Clean Water Act, authorizing the EPA to treat Indian tribes as states under certain circumstances. Through the amendment, Congress granted tribes jurisdiction to regulate their water resources in the same manner as states. Isleta immediately began the process, working closely with the Environmental Protection Agency Region 6.

The Pueblo of Isleta received funding to conduct some model projects that eventually led to water quality standards as strict as drinking water standards. "Our plan and our goal was to establish water quality standards for the Pueblo of Isleta that would exceed, certainly not only the city's water quality standards but also the state's and everyone else's for that matter," Verna Teller said.

While federal arsenic levels are set at 50 parts per billion (ppb), Isleta Pueblo's allowable arsenic levels were set at 17 parts per thousand (ppt). Isleta Pueblo adopted water quality standards for Rio Grande water flowing through the tribal reservation, which were approved by the EPA on December 24, 1992. This strategic move enabled the tribe to negotiate from the highest level of quality possible. The City of Albuquerque was given a grace period of three years to upgrade sewage treatments and analyze discharges.[171]

Verna stated:

Many of the local governments north of us gave us support through resolutions and letters, which we used as we sat and talked with the City of Albuquerque and also with the EPA. It took us 18 months to get William Riley to come to Isleta Pueblo and visit us. We made him take a swim. He finally realized that we were right; we do need to clean up the Rio Grande. He sent a team of people in their white suits to Isleta a few weeks later and they did a soil testing and water testing. We were real happy about that, but it was interesting that we had to go through so much trouble, and it took us a long time to be able to get the man down there.

Verna continued to stress how detrimental poor water quality is to the health of her community and the environment, simultaneously asserting the right to religious freedom. "What we're saying is if we can't drink the water we cannot complete our ceremonies because the water is not clean. If we can't drink it because of all the toxins in it, then our religious freedom is being impinged upon. We cannot complete our ceremonies the way we should. And that is a very serious situation for Indian people."

"Ceremonial use" encompasses water used for either religious or traditional purposes. The policy expressed in the American Indian Religious Freedom Act states, "It shall be the policy of the United States to protect and preserve for American Indians their inherent right of freedom to believe, express, and exercise the traditional religions of the American Indian... including but not limited to...the freedom to worship through ceremonials and traditional rites."

Discharge from the water reclamation plant is controlled by a National Pollution Discharge Elimination System (NPDES) permit. The city's permit was being revised when Isleta's water quality standards were approved by the EPA; as a result, the City of Albuquerque was required to meet Isleta Pueblo's more stringent standards. The city challenged the EPA's approval of Isleta's water quality standards on numerous grounds, further questioning whether

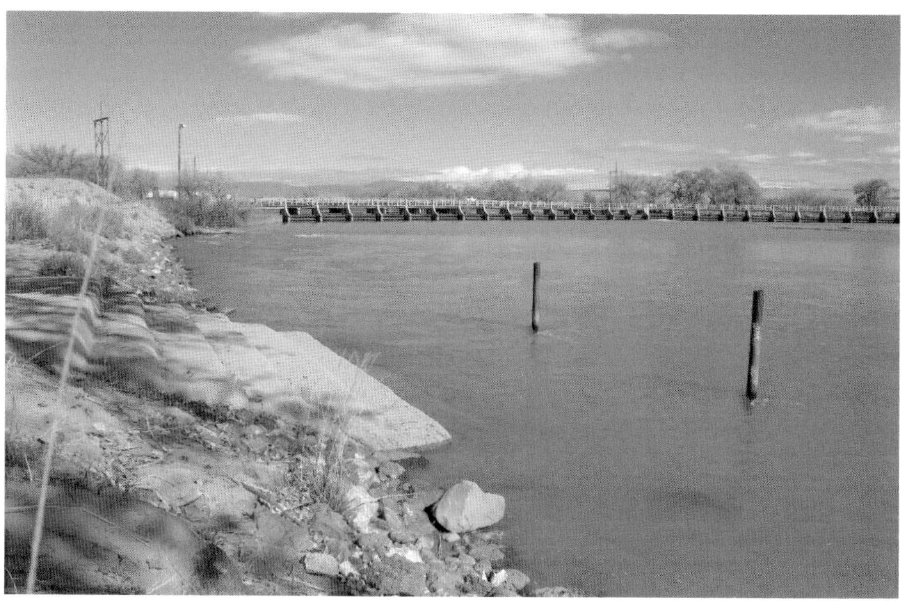

Downstream water from Albuquerque entering Isleta Pueblo, February 7, 2017. *Photo credit: Valerie Rangel.*

the EPA reasonably interpreted the Clean Water Act by approving Isleta Pueblo's standards and whether the EPA's approval of Isleta's ceremonial use designation offends the Establishment Clause of the First Amendment. Implementation of the higher water quality standards meant that the city would need to create a wastewater treatment facility capable of meeting stricter standards, a feat deemed exorbitantly expensive and difficult due to the groundwater's naturally occurring arsenic levels.

On April 15, 1994, the City of Albuquerque, the EPA, the State of New Mexico and Isleta Pueblo agreed to a new four-year NPDES permit for Albuquerque pursuant to a stipulation and agreement. The stipulation and agreement does not mention the claims in this suit, the EPA's regulations and the Isleta Pueblo's revised water quality standards, which are now in effect. During the briefing stage of the appeal, Albuquerque filed a motion requesting an order vacating the district court's judgment due to mootness and remand with instructions to dismiss its complaint without prejudice.[172]

Albuquerque raised seven issues in its appeal: (1) whether the district court's opinion and order should be vacated because the case is mooted by an agreement negotiated by the parties; (2) whether the EPA reasonably interpreted § 1377 of the Clean Water Act as providing the Isleta Pueblo's authority to adopt water quality standards that are more stringent than required by the statute, and whether the Isleta Pueblo standards can be applied by the EPA to upstream permit users; (3) whether the EPA complied with the Administrative Procedure Act's notice and comment requirements in approving the Isleta Pueblo's standards under the Clean Water Act; (4) whether the EPA's approval of the Isleta Pueblo's standards was supported by a rational basis; (5) whether the EPA's adoption of regulations providing for mediation or arbitration to resolve disputes over unreasonable consequences of a tribe's water quality standards is a reasonable interpretation of § 1377(e) of the Clean Water Act; (6) whether the EPA's approval of the Isleta Pueblo's ceremonial use designation offends the Establishment Clause of the First Amendment; and (7) whether the Isleta Pueblo's standards approved by the EPA are so vague as to deprive Albuquerque of due process.[173]

The Supreme Court's decision ruled in favor of the Pueblo of Isleta; the City of Albuquerque was forced to comply with the higher standards set by the tribe. The Pueblo of Isleta inspired many other sovereign nations; seven tribes located along major rivers in New Mexico have developed water quality standards, while many more tribes in the United States have initiated the process.

Ongoing Contamination

On February 27, 2015, the City of Albuquerque experienced issues related to a power outage that resulted in six million gallons of sewage being spilled into the Rio Grande and entering the Pueblo of Isleta. On March 11, 2015, Charles S. Leder, PE, plant operations division manager of Albuquerque Bernalillo County Water Utility Authority, addressed the Isleta Tribal Council concerning the spill: "We apologize that we did not do our job. We were not prepared."[174]

Council member Fernando Abeita questioned why there was a delay in getting notice out to the community about the spillage into the river. He asked Leder why they did not immediately inform the news media so notice could get out. Leder responded, "You bring up a good point about notification and there is room for improvement to allow people to get notice."

Council member Verna Teller couldn't understand why Albuquerque was not prepared. She said, "You should have been prepared." Leder replied that there were issues with a power outage from the Public Service Company of New Mexico; however, the company was "committed to correcting those problems." Teller asked, "Don't you have a backup system?" Leder explained that their internal switching did not perform the way they expected. Teller said, "I'm not satisfied with your explanation. The contaminated water impacted everyone down the river and you should have been here the very next day."

Governor Eddie Paul Torres told the Isleta council to limit its discussion with Leder to just ten minutes. Though the discussion went beyond ten minutes, there was no specific mention of the E. coli that was discovered in the river or whether there would be any long-term or adverse effects to the communities' health or to the health of the river, the vegetation or the fish.

According to Public Health New Mexico's interactive online timeline of Clean Water Act violations, the city's water plant is only one of many nearby sources of pollution affecting the Rio Grande. Rio Rancho's wastewater treatment plant has been in violation of environmental laws, and the City of Albuquerque has at times drained nearly six million times its permitted amount of aluminum into the watershed through stormwater drains along with lead, PCBs and arsenic. The Delta Person Station also has numerous violations for excess chlorine discharge, and Sandia Peak Ski Company was in violation for an E. coli discharge. Over the past decade, more than seven hundred pounds of lead has legally washed into the Rio Grande watershed from Kirtland Air Force Base.[175]

Government records show that over the past decade, facilities upstream from the tribe have been in violation of the Clean Water Act, the federal law that regulates river pollution, for releasing undertreated sewage and for spilling chlorine into the river. Verna Teller said pollution in the river has definitely affected another aspect of the tribe's way of life: the religious ceremonies that take place in the Rio Grande. "Our ceremonies and our religious culture here in our community are of utmost importance to us," she said. "That's what's kept us who we are as a people. And so when there's any threat to that, it's very frightening for our people. It's a threat to our existence as far as we're concerned."[176]

9
SALT WOMAN

Zuni Salt Lake, also known as Zuni Salt Lake, or Fence Lake, is referred to as Áshįįh by Navajos and known as Las Salinas by Hispanic settlers. The area is a neutral zone whose sanctity is shared by various Native American tribes of New Mexico. The Pueblo of Zuni led a historic battle that spanned nearly two decades to protect this sacred site from a proposed coal mine and oil development, which threatened groundwater, natural resources and native rights to freedom of religion and the continuation of traditional beliefs.

Located sixty miles south of the Pueblo of Zuni, in Catron County, New Mexico, Zuni Salt Lake is a shallow saline lake supported by runoff and saline springs. Limnologists estimate that the maar, a low-relief volcanic crater, was formed during the late Pleistocene period, when volcanic and phreatic explosions punctured Paleozoic and Mesozoic geologic strata. During the wet season, the shallow lake carries a depth of four feet, becoming salt flats when water evaporates during the summer. The layer of salt left on the lake bottom has been harvested for centuries by several different tribes within the area.[177]

Ancient roadways radiate from the lake and lead to several pueblos and prehistoric archaeological sites such as Chaco Canyon. Pueblo people of the Southwest have made annual pilgrimages to Zuni Salt Lake and harvested salt for domestic and ceremonial purposes. Acoma Pueblo, for example, cannot hold certain summer dances without the salt from Salt Woman. Use of salt from Zuni Salt Lake is also culturally important to the Diné, who use it to bless the first tear and smile of a newborn. Navajo leader Manuelito said during the signing of the Meriwether Treaty that as long as he and his people could remember, they had journeyed to the salina near Zuni to gather salt. The Salt Lake was a neutral zone where anyone could obtain the precious life-sustaining mineral without harm.[178]

The Salt Lake is especially important to the A:shiwi (Zunis), who consider the lake to be the embodiment of Salt Woman, Ma'l Oyattsik'I, a central deity in the tribe's religion. Zuni Salt Lake and bordering areas are often referred to as the "Sanctuary Zone," or A:shiwi A:wan Ma'k'yay'a dap an'ullapna Dek'ohannan Dehyakya Dehwanne. Burials and religious shrines are contained within this zone. According to traditional Zuni history, Salt Woman (Ma'l Okyattsik'i) lived north of the spring in Black Rock, near the present-day Zuni village of Halona:wa. From 1693 to 1700, the Zunis consolidated their villages into one site in the area of Halona:wa.[179]

Oral recollections of the A:shiwi hold vital ethics such as reverence for nature, consequence of human actions, religious beliefs and societal values that are interwoven into songs, dances and historical accounts. The history of Salt Woman is embedded with instructions for comprehending time, direction, geography and identity. The oral history of the A:shiwi recalls that Salt Woman provided salt to the Zunis for many years until a time when salt was gathered without leaving proper prayer offerings and the area became desecrated by human defilement. Feeling abused and disrespected, Salt Woman abandoned the tribe and salt at the lake was no longer produced. For a long time, the A:shiwi prayed for her return, and when Salt Woman returned, she warned that if the A:shiwi disrespected her again, she would never return.

In 1540, salt from Zuni Salt Lake was a well-known high quality trade commodity by the time Francisco Vásquez de Coronado visited the village (Hawikkuh) of Zuni. Zuni Salt Lake was not part of the Zuni reservation designated by the Zuni Spanish land grant that was signed in 1689. Zuni leaders traveled to Santa Fe and, on August 8, 1850, signed the Pueblo Treaty of agent James S. Calhoun, which promised protection of tribal land and sovereignty.[180]

From 1846 to 1876, Zunis lost control of an estimated nine million acres of territory. During this time, the United States established Fort Wingate and began extensive ranching operations. The first portion of the Zuni Reservation was designated by executive order in 1877. By 1881, railroad tracks ran through the middle of Zuni territory and roads were carved into the mountainsides to facilitate the delivery of lumber harvested from a sawmill in operation in the Zuni Mountains. During the 1890s, more than 200,000 American-owned sheep were herded in the Zuni Mountains, resulting in damage to the upper watershed.

Salt harvested from the lake was an important trade item for the tribe until 1906, when the Zuni tribe was prohibited from gathering salt at the lake and the U.S. government allowed non-Zunis to begin commercial operations at the lake. During the 1930s, Zuni leaders petitioned to have religious areas reincorporated to the reservation; Zuni governor Henry Gaspar took U.S. officials to visit sacred areas and petitioned for land repatriation through the 1940s.

In 1978, an act of Congress provided the return of Zuni Salt Lake to the Zuni Nation and allowed the Zunis to sue the government for lands taken without payment. Zuni filed a land claim suit in the U.S. Court of Claims in 1979, then, in 1985, the Zuni tribe purchased the lake and surrounding land, once again designating the area a "neutral zone" where Spanish, Mexican and American settlers who moved into the region could abandon their differences and gather salt.

The United States allowed the Salt River Project (SRP), the third-largest public utility in the country, to begin mining feasibility tests within the Sanctuary Area for a proposed 18,000-acre Fence Lake Coal Mine. "SRP was first granted a state mining permit to operate in 1996. New Mexico mining regulations require that actual mining on state lease lands begin in the third year of a five-year permit. In 1999, when three years had lapsed, the state permit was extended without Zuni knowledge, and then the state renewed the original five-year permit on July 12, 2001." The site of the proposed coal strip mine was located eleven miles northeast of the lake, on state, federal and privately owned land and was estimated to have mined eighty million tons of coal over a fifty-year time frame, which included a forty-four-mile railroad line to carry the coal to the Coronado Generating Station in St. Johns, Arizona, selling the electricity the coal would have produced to 190,000 residents in Phoenix, Arizona.[181]

While SRP had underground water rights sufficient to extract up to nine hundred gallons per minute, the mine plan called for pumping of eighty-five gallons per minute (primarily to control dust). The Pueblo of Zuni was concerned that pumping water, from the underlying Dakota Aquifer, might reduce the water and salt feeding the Zuni Salt Lake. Following Zuni concerns, the 2002 federal permit required that SRP take water from the Atarque Aquifer, which lies above the Dakota Aquifer. The Pueblo of Zuni argued that depletion of the Atarque Aquifer would also damage the Zuni Salt Lake.[182]

President George H.W. Bush signed Public Law 101-486, the Zuni Land Conservation Act of 1990, which attempted to settle Zuni claims against the United States for damages to Zuni trust lands. In 1996, a state mining permit was granted to SRP. Not only did Native American communities voice their opposition to the mining permit, historical and biological centers also voiced concerns. Jan V. Biella, deputy historic preservation officer for the state of New Mexico, requested that the lake and trails be listed under the National Historic Preservation Act.

In 1999, the National Park Service recognized the significance of the 185,000 acres surrounding the lake, and it was listed in the National Register of Historic Places. Also in 1999, three years after the mining permit was granted to SRP, the company had not complied with the New Mexico state mining regulations requiring actual mining to begin in the third year of a five-year permit on state lease lands. Despite noncompliance and without tribal consultation, the state renewed the original five-year permit on July 12, 2001.

From 1994 to 2003, the Pueblo of Zuni filed several lawsuits against the Fence Lake Coal Strip mine. The proposed construction of the mine and railroad tracks would have not only disturbed ancient trails that many tribes continue to use during pilgrimages today, but coal dust and fine particulates also would have potentially polluted air and surface waters of other sacred sites and communities residing within the area.

While the mine would have generated an estimated two hundred full-time jobs, environmentalists advocated for the preservation of ecological diversity and protection of the Salt Lake's delicate ecosystem. The Center for Biological Diversity (CBD) became a campaigner to save Zuni Salt Lake in the mid-1990s, joining the Pueblo of Zuni in challenging the original state permit granted to operate the mine, objecting to a federal Environmental Impact Statement that concluded "no significant impact" on the Zuni Salt Lake proposed mining project.

Pablo Padilla, a member of the Pueblo of Zuni and the tribe's first director of environmental protection specialist, served the Zuni Council, leading negotiations to retain sacred burial sites and advocating for religious rights. Padilla worked closely with the Zuni Salt Lake Coalition (ZSLC), comprised of members from the Pueblo of Zuni, Center for Biological Diversity, Citizens Coal Council, Sierra Club, Water Information Network, Tonatierra, Seventh Generation Fund and others utilizing grassroots activism, and organizing thousands of supporters to stop the Fence Lake Coal mine. The ZSLC organized a "People's Hearing on Zuni Salt Lake," where more than five hundred people gathered to offer testimony for the protection of Zuni Salt Lake at Zuni Pueblo, holding twenty-four-hour prayer runs around SRP headquarters, marches, and rallies, and engaged media coverage of ZSLC's campaign in the *New York Times*, *Albuquerque Journal*, *Navajo Times*, *Santa Fe New Mexican* and *Arizona Republic*.

In 2001, the Bureau of Indian Affairs released a report by an independent hydrologist who had been studying the potential damage to Salt Lake, which concluded that water pumping associated with mining activity would reduce the water that feeds the salt lake and that the proposed monitoring plan by the company would not be sufficient to detect reductions in water levels of the aquifer. Challenging the hydrologist's report, the Office of Surface Mining recommended renewal of the mining permit.

Although the Center for Biological Diversity and the Pueblo of Zuni disputed the state's renewal of SRP's mining permit, in May 2002, Secretary of the Interior Gale Norton and Deputy Interior Secretary Steven J. Griles, a former lobbyist for the coal industry, approved the mining permit. In the fall of 2002, seven bodies were disturbed during SRP's initial construction of railroad tracks intended to transport coal from the proposed mine. Zunis again voiced concerns about ceremonial shrines and burials in the area. According to the Pueblo of Zuni, an estimated five thousand archaeological sites are located within the mine's lease area, including more than five hundred ancestral human burials.

The Zuni Salt Lake Coalition (ZSLC) continued to campaign, attaining tremendous political support from the New Mexico congressional delegation composed of Senators Peter Domenici (R-NM) and Jeff Bingaman (D-NM), and Representatives Steven Pearce (R-NM) and Tom Udall (D-NM). On July 1, 2002, the congressmen sent a letter to the Department of the Interior and to the Bureau of Indian Affairs, reminding them of the federal government's right to modify or cancel SRP's plan to protect cultural resources and the Zuni Salt Lake, and supporting Zuni Pueblo's request to temporarily suspend SRP's mining activities until pump tests on groundwater near the Zuni Salt Lake could be completed.

In July 2002, Zuni governor Malcolm B. Bowekaty and others testified before the Senate's Committee on Indian Affairs, stating: "Zuni Salt Lake is in real danger of disappearing. The Zuni tribe feels that the U.S. Department of Interior has failed us in its obligation…to protect this sacred lake and associated cultural resources from destruction. Needless to say, the Zuni tribe finds it impossible to rationalize the displacement of our ancestors' burials for the sake of making money."

On the morning of Columbus Day, October 14, 2002, using the tradition of running, members of the Zuni and Hopi tribes undertook a three-hundred-mile run that began at Zuni Pueblo Tribal Headquarters and ended at Papago Park in Phoenix, where an estimated one hundred supporters from the Zuni Salt Lake Coalition were waiting to welcome the runners. The goal of the run was to deliver a message to William Schrader, president of Salt River Project, urging him to stop plans for the Fence Lake Coal Mine so that Zuni Lake and other sacred sites would not be harmed.

The majority of New Mexico's congressional delegation wrote to federal regulators expressing their concern for the lake. On May 29, 2003, Zuni Salt Lake and the Sanctuary Area were listed on the National Trust for Historic Preservation's list of Eleven Most Endangered Historic Places in America. In a surprise turn of events, the Salt River Project announced in a press release on August 4, 2003, that it would relinquish all permits and coal leases for the proposed Fence Lake Coal Strip mine and close down operations; claiming to have found cleaner and more economical sources of coal in the Powder River basin of Wyoming. After the press release, Zuni councilman Dan Simplicio commented, "It's a tremendous victory for all Indian tribes concerned with sacred sites issues."[183]

"We can't risk losing Salt Woman again," Simplicio says. "If we lose her, we might never get her back, and we can't live without her." Simplicio is referring to a Zuni oral tradition that explains how Salt Woman came to Salt Lake. Centuries ago, a moral decline in the Zuni people offended Salt Woman, and she retreated from their lives. The Zunis then entered a period of reflection, prayer and penitence, after which they were shown a new path to Salt Woman. Legend says they followed this new path south for thirty miles and found Salt Lake, where they resumed their worship of Salt Woman. Simplicio says construction of the mine will destroy some of the ancient trails that lead to the lake and potentially disturb as many as five hundred old grave sites. Says Simplicio, "When we were given Salt Lake, we were given an obligation as stewards, caretakers. That is Salt Woman's home, and we have to protect it."[184]

Protests began again in October 2003, when the Bureau of Land Management (BLM) took bids for oil and gas leases on 74 of the 134 parcels of land across thousands of acres within the Sanctuary Zone. The BLM would have to resolve protests filed by a coalition of environmental groups before leases could be issued and drilling permits considered. The National Trust recommended that the BLM designate the entire Sanctuary Zone as an Area of Critical Environmental Concern (ACEC). Under this designation, oil and gas leasing would have been prohibited and groundwater pumping restricted. BLM designated an estimated one-fourth of the Sanctuary Zone as an ACEC, agreeing to develop a Memorandum of Understanding with the Zuni and other tribes establishing a consultation process for how BLM will consult future actions proposed within the ACEC. In February 2004, the Zuni Coalition filed an appeal with the Interior Board of Land Appeals, saying that water pumping for coal-bed methane wells would pose serious risks to the lake.[185]

In 2006, the Interior Board of Land Appeals overturned the leasing decision, ruling that BLM had failed to conduct a meaningful review of cultural resources before lease sales. Zuni Salt Lake remains an undisturbed salt lake, a unique desert oasis, revered for its beauty and natural resources. This place continues to be a vital entity to the Zuni tribe and a sacred place that embraces cultural and religious ceremonies, history and traditions for many other tribes of New Mexico.[186]

According to Pueblo of Zuni tradition, only A:shiwi (Zuni) religious leaders are allowed to visit the area and look upon the Salt Woman Deity. This sacred site is held with such reverence that Zuni tribal members do not gaze upon online images or photographs of Zuni Salt Lake. With highest esteem and respect for the Zuni worldview, no photos of Salt Mother are published in this book.

10
FANTA SE

A community that chooses a dry ditch over a flowing river
is a good place to investigate cultural values.
—*Santa Fe River Ethics*[187]

When founded by Spanish colonists in 1610, the city of Santa Fe's full name was La Villa Real de la Santa Fe de San Francisco de Asís ("the Royal Town of the Holy Faith of St. Francis of Assisi"). Santa Fe means "holy faith" in Spanish, and the city touts that it is the oldest state capital in the nation.[188] Ancestors of the Okay Owingah, Tesuque and Pojoaque Pueblos occupied, farmed and resided in what is present-day downtown starting as early as AD 900 and built a number of villages in the area from 1050 to 1150.[189]

According to Susan Hazen-Hammond, "A Native American group built a cluster of homes that centered around the site of today's Historic Plaza with settlements that spread for half a mile to the south and west; the village was called Ogapoge." Ogapoge, or Ogha Po'oge, means "White Shell Water Place" in Tewa and refers to its sacred water. Navajos called this place Yootó, meaning a string of water beads, thought to visually depict pools of standing water along the Santa Fe River.[190]

Gil Vigil, former governor of Tesuque Pueblo, stated that the founders of the Catholic basilica knew that the pueblo people revered the area of Santa Fe and purposely built the capital city's cathedral over a natural upwelling or spring considered sacred to the pueblo. This action has been perceived by

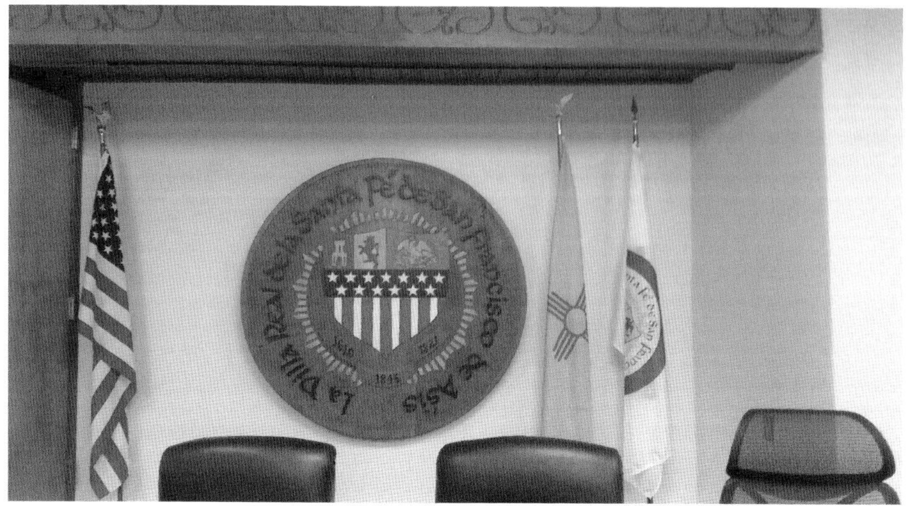

Historic seal of the City of Santa Fe, New Mexico, January 1, 2017. *Photo credit: Valerie Rangel.*

the community as an act of cultural suppression and a tactic of intimidation by the conquistadors.[191] Prior to the building of the cathedral, the original church, built in 1626, was destroyed during the 1680 Pueblo Revolt. An adobe church, La Parroquia, was erected between 1714 and 1717 and then was built over to create the present cathedral basilica erected by Archbishop Jean-Baptiste Lamy from 1869 to 1886.[192]

Historically, pueblo villages irrigated fields with the flow from the Santa Fe River. After the Spanish conquest of the village in 1610, water for agricultural use intensified; more than thirty acequias were established to irrigate an estimated two thousand acres of farmland during the eighteenth and nineteenth centuries. Acequia culture operates under the premise that water is a shared commodity and governed by communal management.[193]

In the nineteenth century, the Santa Fe Trail network and annexation of New Mexico in 1848 brought a new influx of cultures and population increase. In 1880, the railroad imported food, and demand for more water began. In order to secure water for the growing population, the Santa Fe Water Company built its first dam, which held 25 acre-feet of water, approximately two and a half miles upstream from the plaza. By 1904, the old stone dam filled with sediment, as all dams do, and water needs were then met by a larger "Two Mile" Dam just downstream with a storage capacity of 387 acre-feet built in 1893.[194]

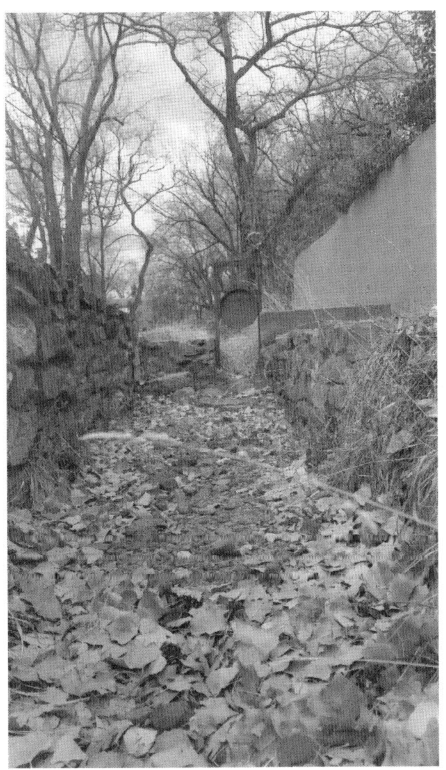

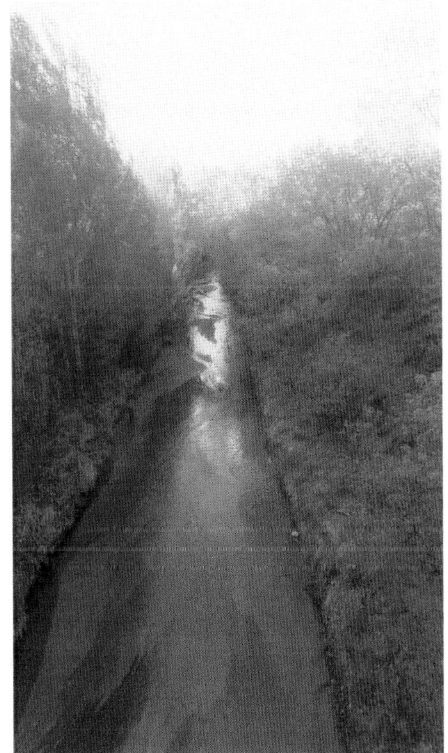

Left: Acequia Madre, Santa Fe, New Mexico, December 24, 2016. *Right*: Santa Fe River, May 15, 2017. *Photo credit: Valerie Rangel.*

In 1919, a survey found twelve hundred acres that were irrigated by thirty-eight ditches within the city and agriculture was the dominant use of water. Another dam was completed five miles upstream in 1928, with double the storage capacity, then a third dam was built in 1943 between the two previous dams. After World War II, erosion of upper watersheds from cattle grazing caused mass soil erosion, exposing bedrock. The braided, meandering river channel would be dramatically altered in the 1970s as the City of Santa Fe tried to contain the river and keep adjacent housing from flooding. The river was reinforced with rock gabions and deliberately downcut through downtown in order to deal with stormwater runoff and acequia channels that dumped water into the river.[195]

Santa Fe's population and tourism industry have exponentially grown, and so has its need for water. The Santa Fe River, in its natural form, is intermittent and ephemeral, flowing after snowmelt and during precipitation events. It experiences several weeks to months of no flow during the year. The river has been dammed, and extraction from wells has caused the riverbed to remain dry for most of the year.

In upstream reaches of the river, dams block the flow in order to provide 45 percent of the city's water supply in normal years. The water is treated in a modern facility near the dams and distributed from there to the city's more than thirty-eight thousand water customers. The remaining 55 percent of the city's water comes from groundwater wells along the Santa Fe River and/or 20 percent from wells just north of town and piped into the city. Nearly the entire flow of the river is owned by the city and stored in surface reservoirs. The city's preferred source supply of water is groundwater, because it is of high quality, is inexpensive to treat and is easy to deliver by gravity.[196]

The city's "River Corridor Master Plan" adopted by council resolution in 1995 called for the release of water for year-round flows in the river, but the policy was never enacted. The city's long-range water supply plan was adopted by city council resolution in 2008 and echoed the 1995 master plan calling for year-round river flows during normal years. The new plan showed how river flow could be accommodated using new sources of water (piped in from the Rio Grande) that would relieve reliance on the Santa Fe reservoirs and groundwater.[197]

Endangered River

In 2007, the Santa Fe River was listed as one of the ten most endangered rivers in the United States. Candidates for the list, "America's Most Endangered Rivers," are nominated by grassroots organizations and are based on the waterway's risk factors, including pollution, water extraction and dams.[198]

Santa Fe's mayor, David Coss, elected in 2006, included the restoration of a "living river" as part of his political platform and established a Santa Fe River Commission to advise the city on how to achieve this goal. The City of Santa Fe Water Division manages releases of surface water from the city reservoirs and supports the "Living River Ordinance." River and watershed coordinator Melissa McDonald said, "Santa Fe River spring flows are timed to provide great benefits for the ecosystems supported by the river, they mimic natural cycles, and are coordinated with acequia irrigation and the Kids' Fishing Derby."

The city's initial spring pulse of 4.5 MG/day occurred from late May through early June. The streamflow then gradually decreased, and the releases of surface water out of Nichols Reservoir was kept at acceptable volume of the total storage available within the reservoir going into the peak of spring runoff season, according to the city. The releases provided a buffer for unexpected increases in streamflow of the Santa Fe River due to continued high temperatures, or rainfall, which could accelerate the rate of snowmelt and streamflow runoff. The city's reservoir management releases water into the Santa Fe River while also allowing snowmelt and spring streamflow runoff to be captured and treated for the city's drinking water supply. Releasing water into the river significantly improves the watershed and riparian ecosystem. In years when the forecast for the runoff from the mountain snowpack falls below 75 percent of the annual average yield, the city will scale back the amount of water released for the "Living Flow."[199]

ENVIRONMENTAL JUSTICE

Historically, the Agua Fria community had large duck ponds located in the riverbed prior to channelization. The river was fed by groundwater upwellings due to the hydrology of an arid climate. Today, the river has rock gabions to control erosion, strategically placed stone waterfalls/riffles to slow down water and concrete and nonnative geology/bedrock lining portions of the riverbed. Upper reaches of the river are located in more affluent neighborhoods, while downstream communities may never see any flow of water in the river at all.

The river stream was once fed by numerous springs and marshes. Groundwater connections to the river are shallow, and water often floods the downtown PERA building across the street from the state capitol roundhouse building. During the late 1800s, a tributary "Rio Chiquito" of the Santa Fe River once fed the garden of Archbishop Jean-Baptiste Lamy near the present-day cathedral. During the seventeenth, eighteenth and nineteenth centuries, Santa Feans fished trout from the Santa Fe River and ice-skated on pools that formed in winter months.

In recent years, the health of the river has fluctuated. The main point sources of pollution are storm drainages that spew untreated human and animal waste, industrial pollutants and contaminants from roadways. Nonpoint sources of pollution include parking lots, farms, backyards that have been sprayed with pesticides/insecticides/herbicides and human waste from homeless residents and animal waste that has led to a high content of E. coli, a type of fecal coliform bacteria that comes from human and animal waste. Nitrates can cause an increase in algae growth. Algae can rob the water of dissolved oxygen and eventually can kill fish and other aquatic life.

Pollution from accidental spills, agricultural runoff and sewer overflows can also change the pH and affect the river's buffering capacity, endangering the river's aquatic life. Temperature changes to the river flow affect many other parameters in water, including the amount of dissolved oxygen available, the types of plants and animals present and the susceptibility of organisms to parasites, pollution and disease. Causes of temperature changes in the river water include weather conditions, tree canopy shading and discharges into the water from urban sources or groundwater inflows.

Prior to colonialism, early Puebloan inhabitants considered Santa Fe and its river and springs sacred, thereby establishing permanent settlements and agriculture in what is present-day downtown Santa Fe. Spanish colonists relied heavily on the springs and flow of the Santa Fe River, so much that when it was cut off during the Pueblo Revolt of 1680, thousands fled the city.

Dispelling the Myth of Fanta Se

The process of gentrification involves the "renovation and revival of deteriorated urban neighborhoods by means of influx of more affluent residents, which results in increased property values and the displacing of lower-income families and small businesses."[200] Spanish colonists took over the pueblo village located in the heart of Santa Fe and forced the relocation of pueblo inhabitants, thereby displacing them to other nearby pueblos and seizing the agricultural lands, acequia systems and viable water resources that were already in place. The intention of the Spanish colonists was to create a capital city by which more affluent residents would bolster trade and business opportunities; thereby property taxes would increase and greater yield would be harnessed from adjacent communities.

The population and social structures rapidly changed as rural village-based people migrated to urban centers in search of jobs and educational opportunities. Santa Fe became an eclectic city with inhabitants from diverse backgrounds and places. Chris Wilson explains, "Despite this rhetoric of social tolerance, and despite the unifying image provided by ubiquitous adobe-colored stucco, the myth of Santa Fe obscures long-standing cultural and class frictions."[201] Coincidently, UrbanDictionary.com states that "people from Albuquerque mock the name for Santa Fe, New Mexico, due to the variety and amount of 'woo woo' practitioners and seekers, nick naming the town Fanta Se."[202]

The truth about the oldest city in the United States is that the area was gentrified and remains segregated; the poorer population separated from the rich, thus creating a cultural and socioeconomic divide among its population. A predominately Anglo population resides downtown, while more impoverished communities of color have been displaced farther south and at the outskirts of town—farthest from the state's capital. The more affluent communities have built mansion-style homes along the road to the ski basin toward the highway route to Taos and in the upper reaches of the mountains as far as infrastructure can be sustained, which has resulted in urban-sprawl housing development. Ironically, a town whose tourism draws heavily from Spanish and Native American heritage possesses neighboring pueblo residents, diverse Native Americans and communities of color that cannot afford to live "in town" or buy equitable housing or rent or lease homes.

The city's most vital resource, the river, is completely dammed for the sole purpose of human consumption. During wet years and depending on snow accumulation in the Sangre de Cristo Mountains, the City of Santa Fe releases water into the river during summer months, a practice that is a mimicking of natural seasonal fluctuations. Without a "living river," invaluable ecological functions cease, groundwater cannot be recharged and water is unable to provide critical wildlife habitat and water for aquatic organisms. Without water, few organisms can survive.

The future use of water, availability of water and concerns for the protection of the ecological integrity of the area will depend on the wishes of the local community and the vested interest of the Pueblo nations who continue to maintain cultural ties to the land and its sacred water entities. "Culture" in Santa Fe is not determined by colonization and urbanism but is celebrated as the dynamic evolving collective wisdom of many groups. This collected wisdom and shared stories of place and traditions will shape the future of the environment as well as of land use, planning and occupation.

II

CHOCOLATE RIVER OF TRASH

Historically, the flow of the Rio Grande through central New Mexico fluctuated drastically in correspondence with weather events and regional saturation. After the 1880s, the practice of logging deforested regions of Colorado and northwestern New Mexico. Runoff from denuded forests filled downstream rivers with silt and heavy sediment load. The middle Rio Grande valley once possessed a river that spread out in a series of braided channels, creating huge sandbars, eddies and pools and replenished groundwater. The water table was shallow and would rise, flooding the valley that contained more than sixty thousand acres of farmland. These frequent flood events created swamplands and deposited alkali and salt minerals, destroying crops and flooding homes and farming villages that relied on acequia irrigation systems utilized for hundreds of years.

The Middle Rio Grande Conservancy District (MRGCD) was created in 1923 in order to provide flood protection from the Rio Grande, drain swamplands and provide irrigation water for adjacent farmlands. By 1935, the conservancy district had built the storage dam at El Vado and diversion dams at Cochiti Pueblo, Angostura, Isleta Pueblo and San Acacia in an effort to manage the Rio Grande's water use. The conservancy also dug seventeen miles of new drainage and irrigation canals, incorporating another 214 miles of existing canal into the system with nearly 200 miles of riverside levees, a system of jetties and checks alongside the river to protect against floods.[203]

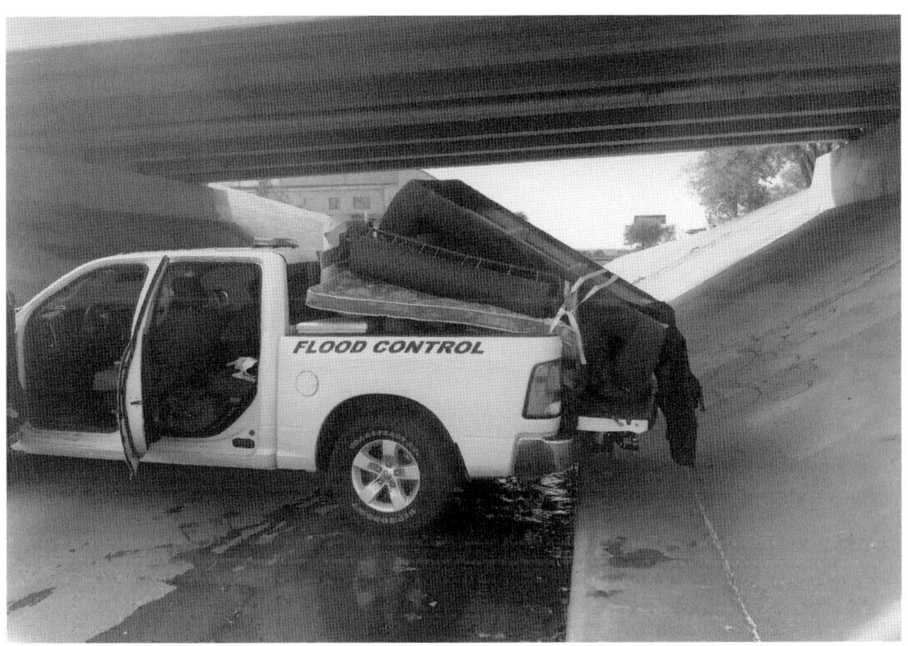

Trash in the Rio Grande. *Photo credit: AMAFCA.*

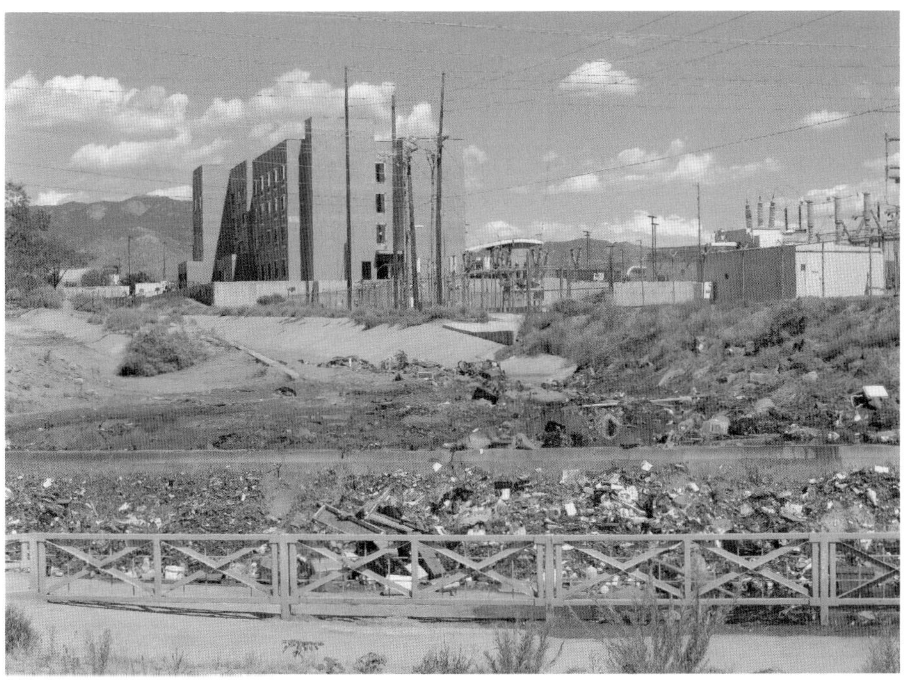

In 1941 and 1942, floods again threatened the middle Rio Grande valley. The MRGCD asked the U.S. Department of the Interior's Bureau of Reclamation for help, and Congress passed the Congressional Flood Control Acts of 1948 and 1950, which provided for the Rio Grande Flood Control Program.

> *The Bureau of Reclamation spent more than $35 million repairing the district's dams and channelizing of 127 miles of the river. The Army Corps of Engineers spent another $40 million on flood control reservoirs and levees, including the construction of a new dam at Cochiti that created the present-day Cochiti Reservoir. The intricate system of ditches, canals, and levees prevents the Rio Grande from overflowing its banks and flowing into the natural braided river system that it once was before, and now only allows for irrigation, agriculture, recreation, and environmental sustainability.*[204]

Within New Mexico, all tributaries east of the Continental Divide drain into the major river, the Rio Grande, which empties into the sea of the Gulf of Mexico. The river spans the distance starting from southern Colorado, through New Mexico, then Texas and down to Mexico. Trash and unseen contaminants traverse the river's waterway, accumulating and compromising water quality for downstream communities along the way. By the time water in the river reaches Mexico, there is often little to no flow.

Water Quality

Color, clarity, smell and taste are indicators of water quality. According to an NPR report in 2013, kayakers in the Rio Grande near Laredo, Texas, found the river to be chocolate brown in color, with a pungent smell, and the water contained floating carcasses of dogs and fish.

In 1992, when Mexico and the United States signed the North American Free Trade Agreement (NAFTA), rapid population growth of border towns ensued, resulting in problems of overpumping of water and polluted runoff with high quantities of E. coli. Chemical waste and raw sewage was dumped into the river south of the Mexican border in the town of Nuevo Laredo. Waste issues continued until a wastewater treatment plant was built in Nuevo Laredo in 1996. The plant was an international waste treatment plant built

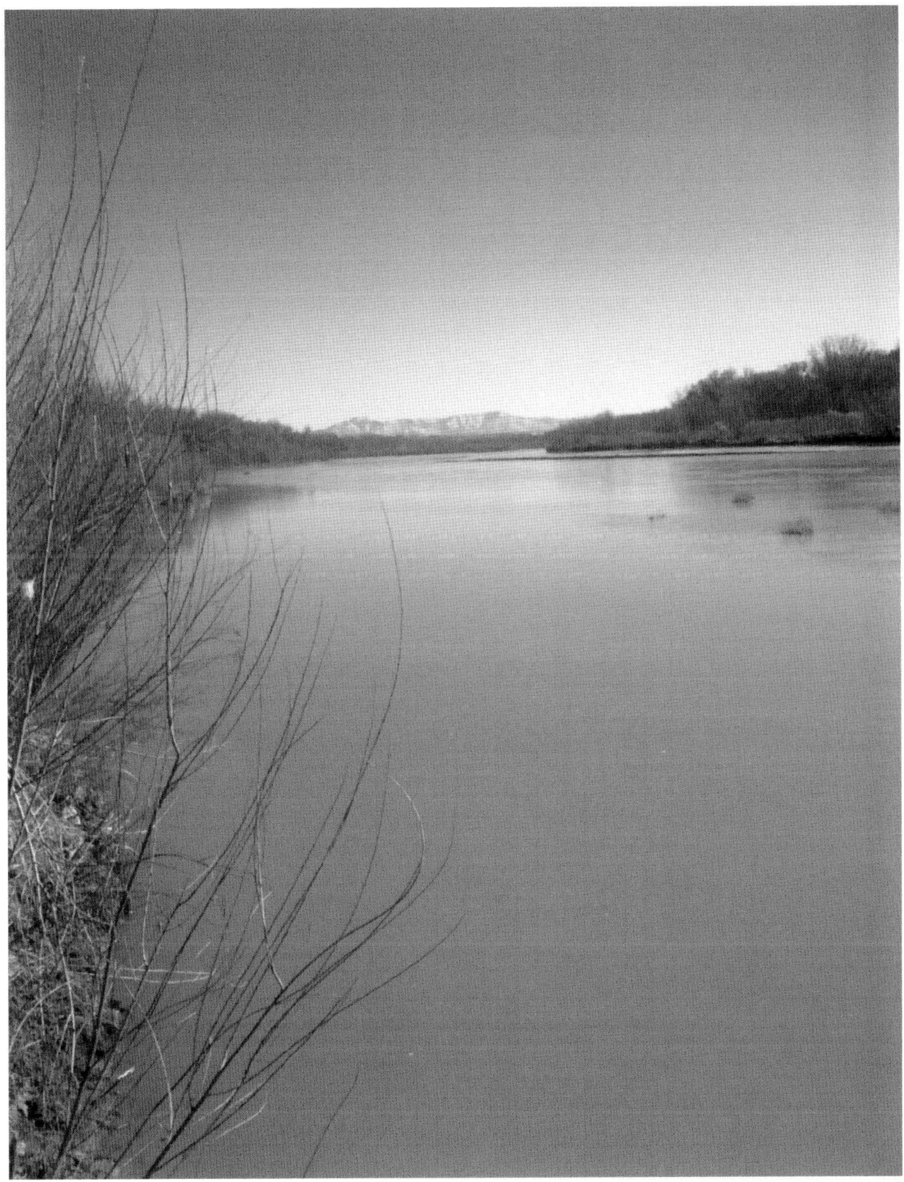

Rio Grande at the Albuquerque Botanical Garden, February 16, 2017. *Photo credit: Valerie Rangel.*

in partnership with both the state of Texas and the country of Mexico. At times, when the water treatment system cannot handle the demands from thousands of residents, the water in Laredo tests positive for E. coli, and water users are told to boil their tap water.

After the NAFTA agreement, a river cleanup was initiated for twelve hundred miles along the border of Texas and Mexico. Water quality is a boundary issue that has existed for decades. The responsibility of river monitoring rests on the U.S. International Boundary and Water Commission, which is part of the U.S. State Department. According to Adolfo Mata, program manager of the Laredo field office, "the Rio Grande was so dirty that people actually died from swimming in it."[205]

In 1994, the Mexican and U.S. governments completed the first phase of the "Bi-national Study Regarding the Presence of Toxic Substances in the Rio Grande/Rio Bravo and its Tributaries Along the Boundary Between the United States and Mexico," an analysis by Dr. Jim Earhart of the EPA, which found that of "thirty toxic chemicals in excess of various governmental screening levels between El Paso and Brownsville, nineteen were in the [Laredo] area. Arsenic levels in a fillet of bass were 11.1 times the Environmental Protection Agency's edible tissue criterion near the Jefferson Street Water Treatment Plant—the sole source of Laredo's drinking water. Levels of copper, zinc and mercury in carp and bass near the drinking water intake were also dangerously high according to safety levels set by the U.S. Fish and Wildlife Service."[206]

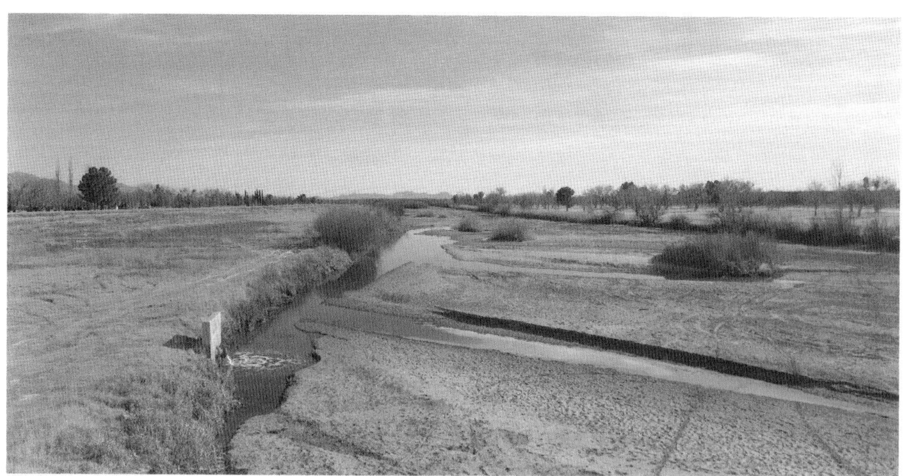

Farm irrigation runoff into the Rio Grande at Anthony, New Mexico (facing south), February 9, 2017. *Photo credit: Valerie Rangel.*

In October 1999, Laredo city officials discovered a "mountain of sludge, concrete, plastic bags, old tires, used sofas and empty industrial containers next to a channel running into the river." Included in the rubble were 75 empty containers of muriatic acid, a chemical used to clean air conditioners and toilets that can be harmful if inhaled.[207] According to a study conducted in January 2000 by the Texas Natural Resources Conservation Commission (TNRCC), pollutants from a warehouse district in northwest Laredo made their way into the Rio Grande. TNRCC found more than twelve hundred warehouses in the Laredo area, many of which are located near tributaries that do not possess protection for spills or other safeguards for preventing contaminants from entering the river.[208]

WATER QUANTITY

The World Wildlife Fund cites inadequate water supply as the main reason that the Rio Grande–Rio Bravo continues to rank as one of the world's ten most endangered rivers. Over-extraction, pollution from trash, raw sewage, industrial waste, over-allocation, invasive species and drought are also issues affecting the river's health.[209] Dams, including weir dams, have cut off water to the main Rio Grande, and the pressures of agriculture compounded by inefficient methods of irrigation account for 80 percent of water use in the United States and Mexico.

More than six million people in several major cities depend on Rio Grande water, and the river water irrigates more than 3,100 square miles of farmland in the United States and Mexico. Water officials have over-allocated for agriculture with unregulated permitting in New Mexico, and an increasing population along the Mexico border only adds to the problem of not having enough water for downstream communities. There are also new demands for water from industrial hydraulic fracturing (fracking).[210]

At times, stretches of the Rio Grande have no flow at all in New Mexico, and flows often don't reach the Gulf of Mexico, where the river water is supposed to empty. WildEarth Guardians, an environmental group, says that the river's health is being compromised. The group wants to free up more water within the system so that it can be used to boost in-stream flows. In March 2016, a lawsuit was filed in the state district court of Santa Fe challenging decades-old permits that allow one of New Mexico's major irrigation districts to pull water from the Rio Grande.

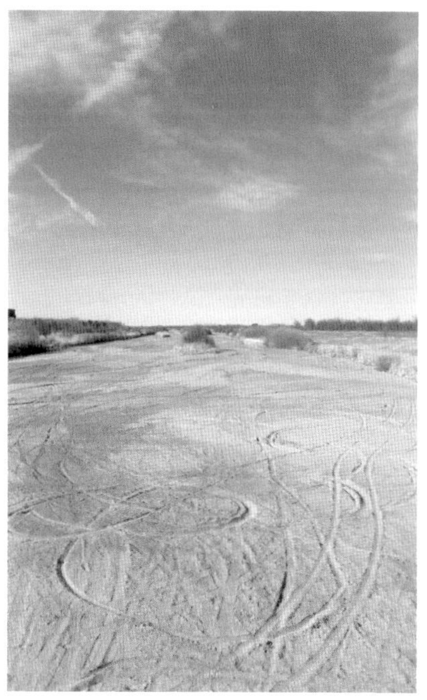

Rio Grande at Anthony, New Mexico, facing north, February 9, 2017. *Photo credit: Valerie Rangel.*

The WildEarth Guardians claim that "the Middle Rio Grande Conservancy District—which delivers irrigation water to more than 100 square miles of cropland in central New Mexico—has failed to prove to the state engineer's office that the water being diverting under permits issued more than 80 years ago is being put to beneficial use." The group is also concerned with the impact of climate change on water availability for municipalities, farmers and endangered species. WildEarth Guardians also claims that the state engineer's office has not required water users to submit proof of beneficial use, blindly allowing permits for water use to continue unregulated or unchecked. "The district does not have any right, let alone an inalienable right, to control the entire flow of the Rio Grande," said Jen Pelz, director of WildEarth Guardians.[211]

In a report released by the U.S. Interior Department, researchers stated, "The reliability of the Rio Grande to meet future needs in the study area is severely compromised by a growing gap between demand and availability and the potential for diminishing supplies due to climate change and competing uses." Invasive species such as giant cane (Carrizo), salt cedar, Guinea Grass and Asiatic clams consume more water than native drought-tolerant species and also drain the river of its water level. During periodic and prolonged drought conditions, water levels of the Rio Grande have been so low that endangered species such as the silvery minnow have not been able to survive and now face extinction.[212]

Water Rights and Adjudication

The adjudication of water rights of the Rio Grande presumably began during the time that ancient Puebloan villages were built along the major rivers in New Mexico. Water was communally cared for and equitably distributed among indigenous populations. Under the 1938 Rio Grande Compact, appropriations were established for the waters of the Rio Grande above Fort Quitman, Texas, among the three states. The compact provides for administration by a commission consisting of the state engineers of Colorado and New Mexico, a commissioner appointed by the governor of Texas, and a federal representative designated by the president of the United States who serves as the chairman.

Presently, the commission meets annually in March to hear reports and consider and adopt an annual report for the previous calendar year. The Rio Grande Compact establishes, among other things, annual water delivery obligations and depletion entitlements for Colorado and New Mexico. Given the variable climate, it provides for debits and credits to be carried over from year to year until extinguished under provisions of the compact. Accrued credits or debits are an important element of compact accounting. The engineer advises and meets with compact commissioners prior to the annual Rio Grande Compact Commission meeting to determine scheduled and actual delivery of water under the compact.[213]

In the 1944 Water Treaty, the United States agreed to give water to Mexico from the Colorado River in exchange for water from the Rio Grande in five-year cycles. Although not clearly defined, the only excuse for not being able to deliver water as stated in the treaty is if the state of New Mexico is in an extraordinary drought. While Texas has consistently honored the agreement, New Mexico has repeatedly failed to deliver the amount of water specified in the treaty. In 2008, Colorado, New Mexico and Texas entered into an agreement detailing an operational plan to better manage and allocate flows in the Rio Grande, recognizing the potential lawsuits over water due to noncompliance. In 2011, New Mexico sued Texas to invalidate the 2008 agreement. Texas then countersued for violation of the compact per the original 1938 conditions. These compact violations are still awaiting judgement by the U.S. Supreme Court. Texas faces problems related to providing enough water to the Rio Grande's users, but these problems arise from the issue of agreement violations by both Mexico and New Mexico.[214]

Water uses of the Rio Grande include agriculture, municipal, industrial and environmental beneficial uses. Both Mexico and New Mexico have over-allocated water to their many users, while Texas is concerned about providing enough water to the Rio Grande's users. The overall issue is the systemic structure of managing water resources by differing political entities. The upper basin stretches through Colorado, New Mexico and Texas and is governed by the 1938 Rio Grande Compact and each state's respective water laws and management strategies, while the lower basin covers a small area of Texas but primarily resides in Mexico.[215]

The majority of issues that Texas faces are driven by climatic factors and management strategies in the Mexican portion of the basin. An announcement was made in the New Mexico Political Report on February 10, 2017, that a lawsuit over the waters of the Rio Grande will head to the U.S. Supreme Court. According to the Associated Press, if Texas were to win the "Rio Grande case," it would have serious implications for some southern New Mexico farmers growing cotton, pecans and chile. Water has been key in the economic development of the state. If farmers, many of whom divert water from the Rio Grande, are forced to stop pumping groundwater, they may not be able to pay for municipal water to irrigate their fields. Should the Supreme Court decide in favor of Texas, New Mexico would be responsible for paying up to $1 billion in damages.[216]

Climate Change

As climate change continues, and temperatures increase within the region and also on a global level, snowpack in mountain regions of New Mexico will decline, rivers and streams will likely have less water in-stream, and higher evaporation will take place in lakes and reservoirs. An interdisciplinary group of university scientists studying drought impacts in the lower Rio Grande region found that groundwater pumping was outpacing underground aquifers' ability to recharge. The scientists noted that the Mesilla Valley aquifer "may no longer have the capacity to provide a reliable, supplemental supply during extended drought conditions and with the current levels of intensive use of groundwater."[217]

Diminishing quality and quantity of water flow in the lower Rio Grande basin are environmental justice issues. This region comprises some of the poorest counties in the United States, with communities that cannot afford to purchase water from other sources, nor can they afford wastewater treatment systems for the polluted river water that exists. Downstream communities, including aquatic organisms, plants and animals, rarely get their allocation of water supply, thus impacting the ecological function that the riparian system naturally provides.

Water management must take into consideration the following: regional population growth patterns, land use planning practices, increased water demand and climate change in regard to the availability of water supplies based on environmental conditions. Surface-groundwater interactions, aquifer capacity and recharge zones, water quality and ecological functions are also critical to the effective management and environmental equity of the Rio Grande basin.

PART III

LAND CONTAMINATION

12

WHAT'S IN YOUR BACKYARD?

In every major city across the United States, there are instances of environmental injustice. Contamination finds its way into marginalized communities of color, and it seems that the brunt of human waste and industrial generation winds up in the poor community's backyard rather than in the most affluent neighborhood. Does the average American really know what's in their backyard, buried in the soil, down the stream or floating in the aquifer below?

Are people aware of their impact on downstream communities? Are they knowledgeable of the contents of the air in which they breathe? Do they realize their place and proximity to industrial zones and resource extraction sites? Do you know what's lurking around the corner, leaking down the block, buried on the other side of town, dumped at the bottom of the hill or tossed in the arroyo?

The designation of "Superfund" means that a site is so polluted that the federal government has determined the place to be harmful to public health or the environment and in need of immediate cleanup efforts. In 1980, the CERCLA federal law was established, authorizing the U.S. Environmental Protection Agency (EPA) to create a list of polluted locations requiring a long-term response to cleanup of hazardous material contaminations. These designated locations are known as "Superfund sites" and were first placed on a Hazard Ranking System (HRS) in 1981, then on a National Priorities List (NPL) in 1982.[218]

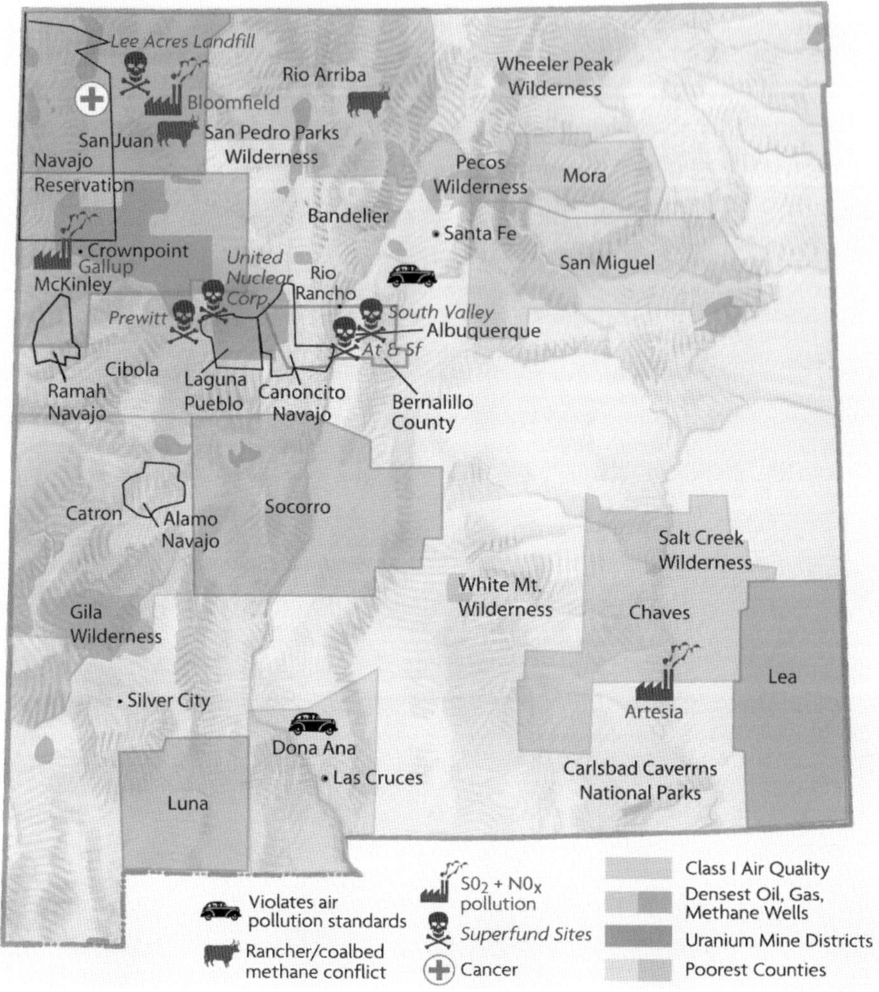

Dreaming New Mexico map. www.dreamingnewmexico.org/energy/environmental-justice. *Photo credit: Bioneers.*

Based on the EPA's most current list for New Mexico, Superfund sites are located in the following fourteen cities: Albuquerque, Roswell, Church Rock, Prewitt, Espanola, Farmington, Laguna Pueblo, Milan, Las Cruces, Socorro, Grants, Silver, Carrizozo and Questa.[219] Only the town of Farmington is not a marginalized community of color, possessing a population of 49.5 percent white ethnicity and with the average income exceeding the state's average.[220]

The list of Superfund sites located in New Mexico reveals a story of resource extraction and dumping in poor neighborhoods, where the workforce is largely low-income populations of Hispanic or American Indian heritage. The most heavily mined and extracted sites of uranium have occurred on or immediately adjacent to land bordering American Indian tribal lands. Four major pollution sites are located in the Pueblo of Laguna and towns of Church Rock, Milan and Farmington. All of these towns have ongoing remediation issues that will likely never be fully resolved, due to the nature of the contaminants.

Questa and Silver City bear the unfortunate legacy of mining with populations struggling to survive economically. Companies utilized the local, cheap labor force for resource extraction until the techniques for extracting minerals and energy production became too costly. When the market was unable to compensate for the materials that were stripped of the earth, workers moved to more lucrative towns, and mining companies closed operations, leaving behind large volumes of waste.

HEALTH CONCERNS

Immediate concerns for the cities that host Superfund sites in their backyards are the health impacts to the residents therein, as well as the future planning and designated use of the contaminated environment impacted. Often, pump-and-treat systems cannot remedy the situation fully, or treatment spreads toxic plumes farther into the groundwater rather than absolve them. Over time, low-level exposure to evaporating chemicals, vapor fumes, tailings dust and tainted well water will likely affect human health, depending on each person's tolerance and other factors such as lifestyle and genetics.

There are an unknown number of buried waste and illegal dumping sites in arroyos, as well as the known state permits that allow private companies and industries to discharge waste into natural waterways. Non-point source pollutants such as nitrates, bacteria, pesticides and fertilizers from farms; chemicals spilled on the ground or from individuals' backyards; animal waste; and byproducts from ranching industries all find their way through cracks and crevices, seeping into bedrock. Many contaminates penetrate through soils, are not digested by plants or eliminated by wastewater treatment and infiltrate deep into the underlying aquifers and then find their way back to the surface through municipal pumping and natural springs.

Superfund sites in New Mexico are a testament to the history of industrial growth and economic booms. Places that once showed prosperity

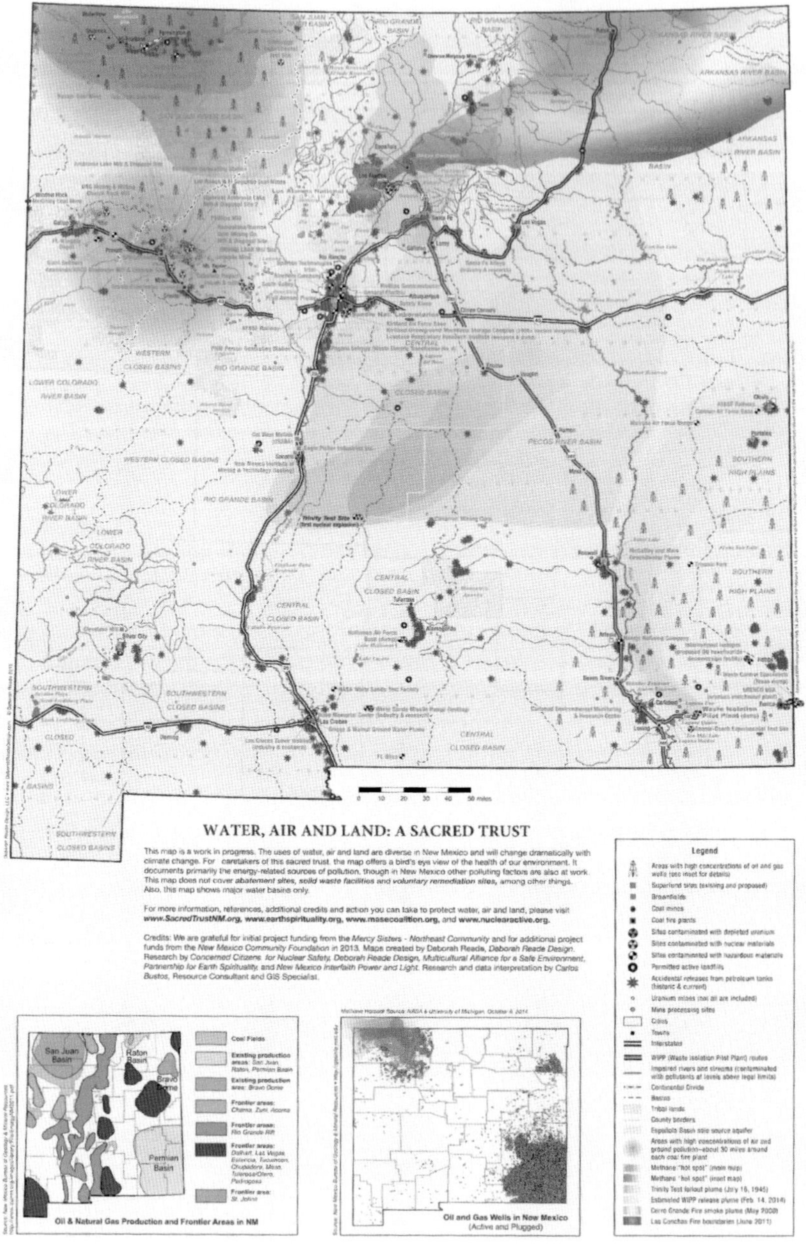

Map of energy-related sources of pollution and major water basins in New Mexico. *Photo credit: Bioneers.*

and promise of economic security are now contaminated lands, remote, struggling communities of color with health disparities. The fact that these sites persist in the contamination that has plagued them emphasizes the stark reality of environmental racism, in which poor communities of color harbor more contamination than affluent cities and neighborhoods throughout the state.

13

MY OTHER CAR IS IN THE ARROYO

Bumper stickers with the phrase "My other car is in the arroyo" are often spotted while driving around the state. Despite the intended humor, New Mexicans know the reality of illegal dumping in arroyos and rivers. With curbside garbage pickup, recycling bins located at convenience centers and designated landfill sites where litter is supposed to be dumped, why on earth would refuse wind up in arroyos and rivers and at the bottom of dams?

The disposal of any material in a place other than the appropriate receptacle is considered "illegal dumping" or "littering." While some materials that have been illegally dumped are items that can be recycled or reused and can easily be gotten rid of by a simple donation to a local Goodwill, Salvation Army, donation centers or sold via Craigslist or through garage sales, the most common items found in arroyos are appliances and electronics that are irreparably broken and cannot be reused.

The City of Santa Fe, for example, has a curbside service in which individuals can call to schedule a pickup for items such as recyclables, electronics, hazardous waste, bulky metals and/or appliances that are not suited for reuse. Depending on the size and quantity of materials, a small fee is charged for this curbside pickup at individuals' residences; large items are picked up at a rate of $25.75 plus tax. Illegal dumping in arroyos and open spaces occurs when city or county residents cannot or do not want to pay to properly dispose of their trash. Trash piles accumulate and pollute the surrounding habitat in which animals and humans interact. City residents, however, are not the only offenders to illegally dump trash in New Mexico.[221]

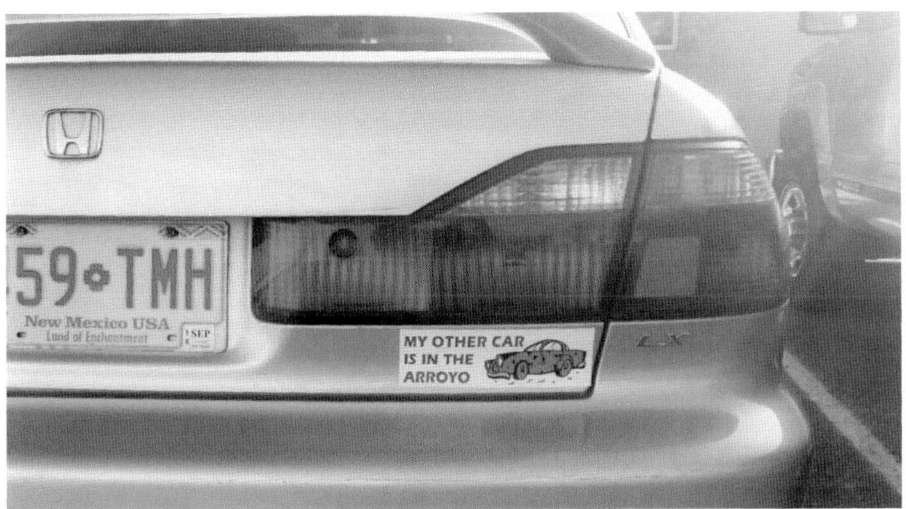

Bumper sticker. *Photo credit: Valerie Rangel.*

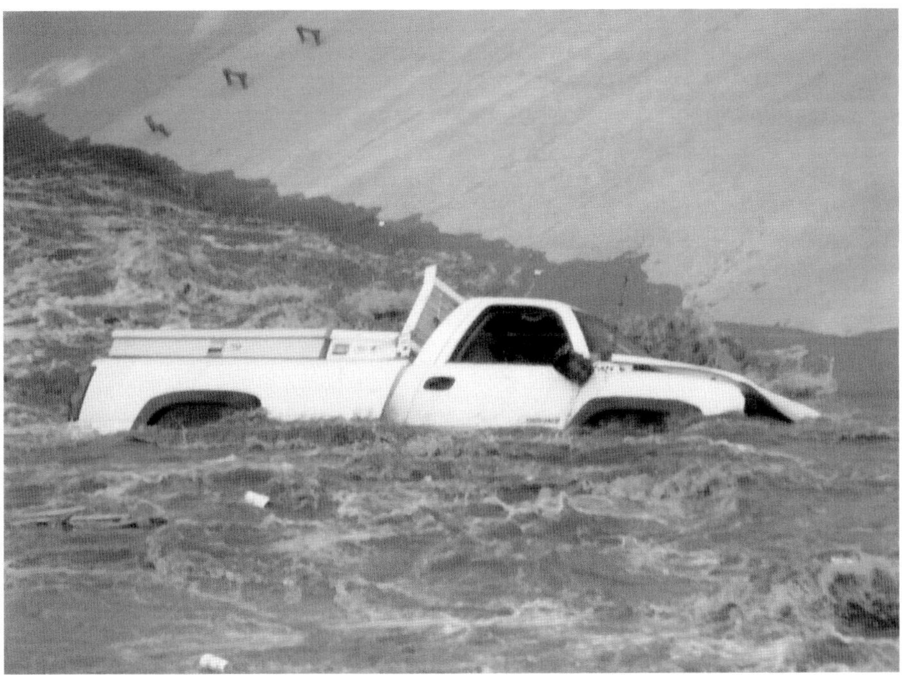

Truck floating in Albuquerque concrete-lined channel. *Photo credit: AMAFCA.*

In 1983, due to a poorly crafted game, the manufacturer of Atari games claimed a loss of more than $300 million and reported that it had $10 million of unsold Atari inventory sitting in warehouses around the country. In March 1983, Atari fired seventeen hundred members of its staff; in April, the company moved its manufacturing plants overseas, closing its El Paso, Texas plant. It is believed that in the summer of 1983, the company chose to cut its losses and illegally dumped an estimated twenty-six hundred cartridges along with *E.T.* paraphernalia into a landfill in Alamogordo, New Mexico, approximately ninety miles from the Texas plant.[222]

Rez Dogs

The saying "One man's trash is another's treasure" refers to what is deemed "worthy" and "unworthy" by an individual, based on personal preferences. While someone might think a fluffy cute puppy for Christmas is the perfect present, another might consider the animal to be an unbearable financial burden and nuisance. Although dumping animals is considered a crime in New Mexico, many out-of-towners driving through the state dump their unwanted animals in remote areas.

According to a letter dated August 7, 2011, written to the editor of the local Sierra County newspaper, "Animal dumping is a huge problem in Sierra County. It is also a fourth degree felony in New Mexico. The Sierra County Humane Society will pursue prosecution provided we have adequate information, i.e: license plate number, etc. Just the other day, a man fishing on the banks of the Rio Grande witnessed an old dark blue van stop quickly by the side of the country road along the river and literally dump 'something' before speeding away."[223]

Dumping of unwanted animals also occurs beyond cities and rural areas of the state, crossing into tribal jurisdiction and onto Indian reservation lands. In 1986, a study of dog bites on the Navajo Reservation found that 60 to 75 percent of all families in urban areas provided shelter or food to at least one dog, while families in more isolated sheep camp areas commonly possessed five dogs.[224]

There are an unknown number of animals dumped in communities near or on reservation lands by people unable to house and care for their pet. Dumping of animals is often the result of the pet becoming unwanted or having a litter of puppies or kittens. Due to a cultural belief that dogs also play an important role in Native American culture and are sacred beings that belong to the spirits, there is a stigma around killing or interfering with their life cycle. Some Navajo families teach their children not to cuddle a dog or bury one when it dies. Dogs traditionally do not enter the home and are expected to live outside, often without shelter from heat and cold. These animals seek food at tourist motels and gas stations and shelter in abandoned buildings or refuse areas.[225]

Poverty-stricken families are unable to provide veterinary care, including vaccinations and spaying and neutering. Communities experiencing high rates of poverty cannot afford added health-care costs or emergency room visits due to attacks and bites from feral dogs. Animal Humane advocates are also concerned with people killing or beating free-roaming dogs. The Navajo Nation Animal Control Program employs a handful of officers to operate and manage four shelters located in Tuba City, Fort Defiance and Many Farms in Arizona and Shiprock in New Mexico. The Hopi Tribal Council in Arizona has finalized an Animal Control Ordinance and is concerned about aggressive dogs and the transmission of diseases. An estimated 140 cases of dog bites occur on the Hopi Reservation each year.[226]

Free-roaming or feral animals have bitten or attacked humans and livestock, spreading diseases such as distemper, parvo and Rocky Mountain spotted fever likely spread by ticks. Some dogs have been hit by vehicles, causing concern for disease transmission to predatory animals such as coyotes and hawks. An estimated three thousand people are treated each year on the Navajo Nation for animal bites and attacks, although those numbers may be low due to the fact that medical treatment may not always be sought or because of underreporting.[227]

KILLING CONTESTS

Not only are unwanted and uncared-for pets "dumped" in remote areas, so also are the carcasses of animals that have been killed for sport or contests. Since livestock operations began in New Mexico, coyotes have been considered a threat and nuisance to landowners. In 2015, nearly forty dead coyotes were discovered in Las Cruces, New Mexico. Some of the coyotes had wood blocks in their mouths marked with the date that they were killed, indicating that they were most likely part of a coyote-killing contest. Though such contests are legal in New Mexico, they are unregulated. In

2014, wildlife advocates counted twenty contests around New Mexico. This number is thought to be low and not representative of all the contests within the state, especially those that are not publicized.[228]

Executive director Kevin Bixby of the Southwest Environmental Center in Las Cruces says, "The animals are not being eaten or used in any way; they are just being killed and they are being killed for sport. This kind of predator killing disrupts natural ecosystems and undermines the ability of coyotes to provide their ecological role in maintaining healthy systems, in regulating populations of prey animals like rodents and rabbits." Wildlife advocacy groups support legislation that would ban coyote-killing contests for material gain but would not prohibit killing coyotes that threaten property, such as livestock or pets. Guy Dicharry of the Los Lunas–based Wildlife Conservation and Advocacy Southwest said, "They are commercial events: killing animals for the purpose of entertainment, prizes and publicity. You're really out there trying to win. This is not focused on predator management. It's random."[229]

In Valencia County, near the state prison in Los Chavez, residents of a small organic farm say that a road was turned into an illegal dump site with large volumes of trash and dead animals, causing a horrible stench. For those caught illegally dumping in Valencia County, fines can be up to $300 or ninety days in prison. Weekly curbside pickup doesn't exist in rural areas, and people must pay $5 for every truckload of trash brought in to the Conejo Transfer Service station. A planned service to pick up trash got stuck in legal battles for years; a solution has yet to appear.[230]

Backyard Junkyards

Dumping of hazardous waste and scrap materials is of great environmental consequence. Individuals are required to possess a permit to house and dispose of hazardous waste and scrap materials. First reported in 2010, in the small town of San Rafael south of Grants, New Mexico, a private property was found to contain hazardous waste that included 250,000 tires, junk cars, scrap metal, trucks, heavy equipment, boats, barrels, asbestos-laden material and piles of old junk appliances. According to the New Mexico Environment Department, there were no permits issued to the private property owner nor permission obtained to house and store these mounds of waste items.[231]

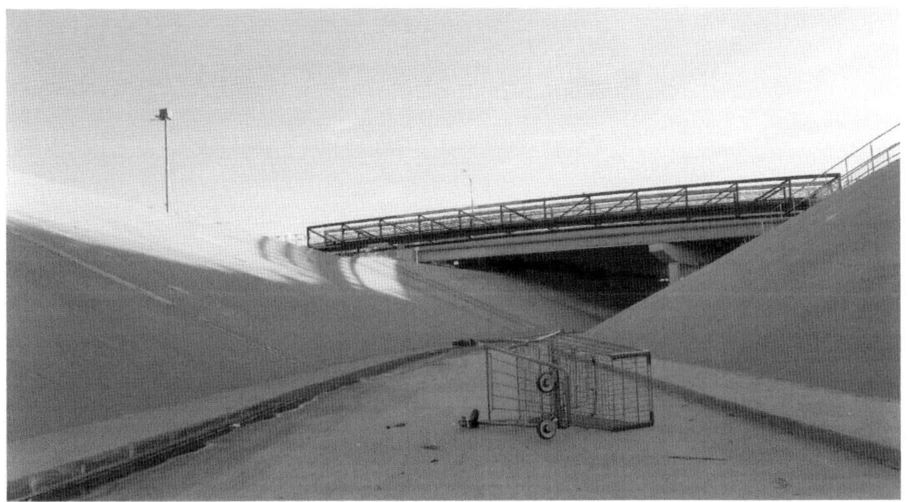

Shopping cart in the concrete-lined channel at Menaul Street, Albuquerque. *Photo credit: Valerie Rangel.*

When the state inspected the property in 2010, it was clear that the property owners had been accepting refuse materials for many years. State regulators promised that the site would be cleaned; however, six years later, the property had not been cleaned, and the owner of the property has yet to be held accountable for the illegal dumpsite. There are concerns that a potential fire could spread hazardous contaminants through the air and that the lack of water resources is insufficient to contain such a fire given the amount of fuel materials on the property. The site poses additional health risks; tires may accumulate water and become a breeding ground for insects that have the potential to spread diseases.[232]

ARROYO DUMPING

According to KRQE News 13 reporter Alex Goldsmith, illegal dumping was on the rise in Albuquerque arroyos in August 2014. Old televisions were the most commonly dumped item within the Albuquerque dams and arroyos, with an increase in shopping carts, presumably dumped by homeless people camping under bridges around arroyos. According to Goldsmith, within one or two days, as many as thirty to thirty-five grocery carts wind up in the channel on the streets of Menaul and Carlisle.[233]

The Albuquerque Metropolitan Arroyo Flood Control Authority (AMAFCA) was created in 1963 by the New Mexico Legislature with specific responsibility for flooding problems in the Greater Albuquerque area. AMAFCA's purpose is to prevent injury or loss of life and to eliminate or minimize property damage. AMAFCA does this by building and maintaining flood-control structures that help alleviate flooding. There are only a dozen AMAFCA maintenance workers to ensure seventy miles of river channels and thirty-five dams are free of debris. Extracting grocery carts out of the arroyo is a challenging endeavor; channels are often steep and difficult to access even with maintenance equipment—especially during the monsoon season, when floodwaters can quickly lift and carry large objects downstream in a matter of minutes.

Social Changes

The tributaries in Rio Arriba and Taos Counties contribute water to the state's two major rivers. Arroyos in these counties often contain discarded mattresses, broken furniture and appliances, old refrigerators, decaying animals and bags of everyday household garbage. Historically, people would burn scrap paper and cardboard material in metal barrels within their backyards; food scraps were fed to chickens, dogs and farm animals; and clothing and tools were repurposed until they had worn out their use.[234]

Some environmentalists say that dumping in rural areas is a side-effect of life in economically depressed areas, including reservation lands where there is an older demographic who are still adjusting from old habits while being overwhelmed by large quantities and different contemporary kinds of garbage.

Perhaps people just don't like to pay to throw things away? Rio Arriba County happens to have one of the most expensive trash systems in the state; many residents are not paying for their door-to-door trash pickup service or are delinquent in payment. Without a nearby landfill, Rio Arriba County hauls its trash eighty-four miles to Rio Rancho, which is why the service is so expensive.[235]

In Taos County, household waste fluids, hazardous and toxic chemicals, biological waste, oil and hydrocarbons gradually leach out of trash, and it all flows into the watersheds that drain into the Rio Grande. Researchers have measured high concentrations of polychlorinated biphenyls (PCBs), which have been consistently high in the tissue of fish at the Taos Junction Bridge near the southern border of Taos County at the point where the Rio Grande flows downstream after intersecting with many of its tributaries.[236]

Bruce Thomson, director of the University of New Mexico's Water Resources Program, said that the improper disposal of household hazardous waste (paint, pesticides, herbicides, automotive fluids and oils) also presents a severe water quality problem. Meanwhile, plastics and other nonbiodegradable materials clog waterways and reservoirs. Asbestos and other chemicals seep from construction waste. Rotten food and animal carcasses contribute bacteria, such as salmonella, downstream. Thomson said, "Whether it's an animal body or yard waste, as it degrades, the bacteria consume the oxygen in the water, and you end up with a body of water that's anaerobic, without any oxygen. Fish kills come from that. In Albuquerque, there are a couple of places where stormwater accumulates, detritus settles out, and we occasionally have anaerobic conditions." These water quality issues are of great concern as the Rio Grande River water is a source of drinking water for many downstream communities as well as the municipalities of Santa Fe and Albuquerque; the river is also a source for irrigation across the state.[237]

Acequia Cultura

In the past, arroyos were maintained by Acequia associations; mayradormos would organize a cleaning seasonally, which would be followed by a huge feast. In September 2008, the Mayordomo Project was spearheaded by a collaboration between the New Mexico Acequia Association and the UNM Ortiz Center for Intercultural Studies. Its purpose was to address a situation that the New Mexico Acequia Association calls the "mayordomo crisis," involving loss of traditional knowledge, attrition and inadequate replacement of New Mexico's mayordomos de las acequias. The mayordomo, or ditch boss, is essential to the practice of acequia irrigation.[238]

According to the New Mexico Acequia Association,

> *Today the old mayordomos are dying and taking their knowledge with them, and they are not being replaced by younger individuals who know how to manage and deliver the water and maintain the ditches in a given community. The Mayordomo Project seeks to investigate and record*

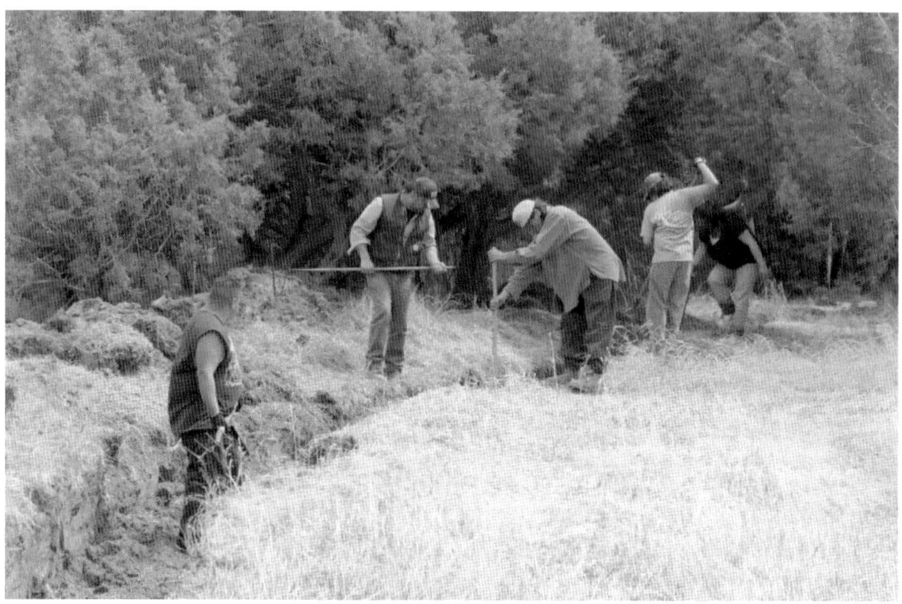

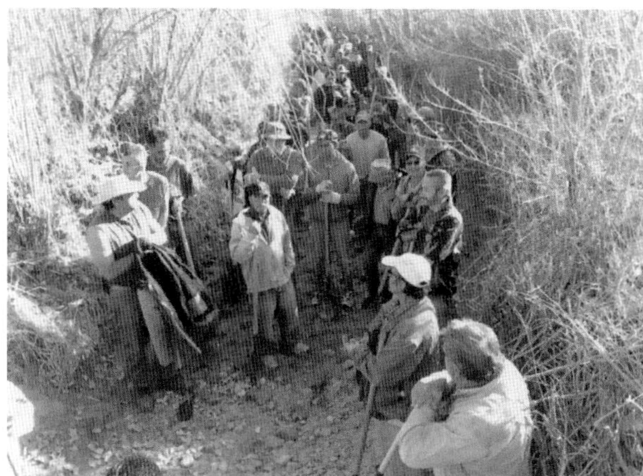

Above: "Team work." *Photo credit: Karen Gonzales, JJ Santistevan, Acequia in Amalia, New Mexico.*

Left: "200 Strong." *Photo credit: Lucia Sanches, Acequia Madre del Alcalde, Alcalde, New Mexico.*

the practical local knowledge of living mayordomos in order to develop a method and program for the transmission of this knowledge to a new generation of mayordomos. It aims to identify, describe, understand and transmit knowledge that is common among all mayordomos as well as particular to a specific individual and location. The methodology of the project is community-based participatory action research (PAR), whereby a community of interest defines a problem it faces and seeks to solve through a collaborative, group process of investigation and action.[239]

While there are great efforts to restore and revive the mayordomía and acequia traditions of the past, today, public land managers rely on volunteers for the cleanup of illegal dump sites. Anti-littering signs and presentations in local schools about recycling and the importance of disposing of trash properly are ways in which Taos and Rio Arriba Counties combat the issue of illegal dumping. The state also reinforces anti-littering through the New Mexico Department of Tourism's Clean & Beautiful program, which partners with affiliated communities, other municipalities and tribal governments throughout our state to beautify New Mexico by means of litter reduction, recycling initiatives and education.

The Clean Water Rule, a major regulation published by the Environmental Protection Agency and the Army Corps of Engineers in 2015, adds to the beautification of the state with promises of enforcing higher penalties to protect intermittent and ephemeral streams that run dry much of the year. Therefore, people who get caught dumping face federal fines in addition to local fines. Many communities across New Mexico have built and maintained public spaces, removed millions of pounds of trash, recycled waste through volunteerism and conducted youth education initiatives to keep the Land of Enchantment beautiful.

14
BRACEROS DE NUEVO MEXICO

The plantation and the ghetto were created by those who had power, both to confine those who had no power and to perpetuate their powerlessness.
—Reverend Martin Luther King Jr.

The history of migrant workers is a history of exploitation. "Migrant farmworker" is defined in the Migrant and Seasonal Agricultural Worker Protection Act (MSPA) as "an individual who is employed in agricultural employment of a seasonal or temporary nature and who is required to be absent overnight from his permanent place of residence." In the early twenty-first century, the U.S. Department of Labor's National Agricultural Worker Survey reported that 77 percent of all workers on American crop farms had been born abroad, and that percentage is almost half of all foreign-born workers living in the United States for less than five years. More than half of these populations of foreign-born workers are not legally authorized to work in the United States.[240]

Most migrant workers are seasonal agricultural workers, many of whom are temporary immigrants who travel to and from the United States from other countries. During the period of Spanish colonialism, priests and friars enslaved indigenous communities, forcing practices of building construction, farm labor and personal indentured servitude. Beginning in the late 1800s, large volumes of immigrant workers were used by the United States to build railroads and infrastructure and to cultivate food. During the 1870s, when slavery was abolished, farmers had to start paying workers to tend their plantations, and a quest for cheap labor began. Chinese immigrants willing to take cheaper wages, as they had little recourse at the time, helped construct the transcontinental railroad. These workers were paid low wages, which kept costs down and increased revenue of produced goods for commercial farms. After the Chinese Exclusion Act of 1882, American farmers employed Japanese workers in the farms of California until the Gentlemen's Agreement of 1907, which stopped Japanese immigration.[241]

In 1910, Punjabi immigrants were the next cheap source of labor, then, during World War I, commercial farms persuaded the U.S. government to exempt Mexicans from restrictions imposed by the Immigration Act of 1917 to slow European immigration. On May 23, 1917, the U.S. Department of Labor allowed this exception, stating, "to admit temporarily otherwise inadmissible aliens." This effectively allowed workers to be employed in agriculture and on railroads and effectively began the first "bracero" program. The Spanish term *bracero* means "manual laborer" or "one who works using his arms." Under this program, almost eighty thousand Sonoran Mexican farmworkers came to the United States prior to 1921 to work primarily cotton and sugar beet fields, even after the establishment of the U.S. Border Patrol in 1923.

Due to the economic impact of the Great Depression, combined with prolonged drought conditions and debilitating haboobs that plagued the Dust Bowel region, many Mexican farmers were sent back to Mexico in order to open jobs for American workers. New labor laws and land reforms were established during this time, and the U.S. government began marketing the idea of a new American dream, to "Move out West" to the "land of sunshine and grapes." Industrial irrigation transformed swamp valleys in the western states into large-scale commercial farms that continue to produce fruits and vegetables for the United States today.

In 1942, as the United States entered World War II, the pressures of a population at war gave rise to a new bracero program that added more immigrant farmworkers, employing almost 500,000 Mexican workers a year during the 1950s. By May 1954, "Operation Wetback" was implemented as an immigration law enforcement initiative created by Joseph Swing, the director of the U.S. Immigration and Naturalization Service (INS), and in cooperation with the Mexican government.[242] "The program is said to have originated from a request by the Mexican government to stop the illegal entry of Mexican laborers into the United States, and was primarily a response to pressure from a broad coalition of farmers and business interests concerned with the effects of Mexican immigrants living in the United States without legal permission."[243]

"After the implementation of 'Operation Wetback,' there was an increase in arrests and deportations by the U.S. Border Patrol; these arrests were deemed 'civil rights violations' which resulted in several hundred United States citizens being illegally deported without being given a chance to prove their citizenship."[244] Nearly five million rural Mexican workers were legally authorized to work in the United States during a second bracero program between 1942 and 1964.

The Bracero Program was controversial in its time.

> *During the 1940s through mid 1950s, Mexican nationals desperate for work were willing to take arduous jobs at wages scorned by most Americans. Farmworkers already living in the United States worried that braceros would compete for jobs and wages would be lowered. In theory, the Bracero Program had safeguards to protect both Mexican and domestic workers, for example, guaranteed payment of at least the prevailing area wage received by native workers. Employment for braceros meant three-fourths of the contract period would provide adequate, sanitary, free housing, and decent meals at reasonable prices, occupational insurance at employer's expense, plus free transportation back to Mexico at the end of the contract. Employers were supposed to hire braceros only in areas of certified domestic labor shortage, and were not to use them as strikebreakers. Farm owners ignored many of these rules and Mexican and U.S. workers suffered while farm wages dropped due to the use of braceros and undocumented laborers who lacked full rights in American society.*[245]

In the late 1950s and '60s, migrant farmers in the West were also hired to tend cattle operations in addition to farm work. These workers were forced to flea-dip animals and apply pesticides and fertilizers to agricultural fields without personal protective gear and with no knowledge of the health risks involved with handling such chemicals. After the publication of Rachel Carson's book *Silent Spring* in 1962, which exposed the detrimental effects of pesticide residues on the natural environment, the public became outraged. During the 1960s, researchers began developing an "integrated pest management" (IPM) program.

In 1964, the U.S. Congress ended the bracero program just as Mexican American farmworker César Chávez and Dolores Huerta created the largest farmworker union to lead a civil rights movement to increase wages and reform conditions for grape harvesters in California. Meanwhile, in New Mexico in the 1960s and '70s, Reies López Tijerina led a land grant movement in an effort to restore New Mexican land grants to the descendants of their Spanish colonial and Mexican owners. In June 1956, Tijerina visited a community in Monero, New Mexico, where he learned of the dispossession of land grants because Hispanic families lacked land titles. Tijerina traveled to Mexico to study the "Laws of the Indies," which had governed the American portion of the Spanish Empire for more than three hundred years. There, he discovered a re-drafted version of the Treaty of Guadalupe Hidalgo containing a protocol that guaranteed land grants to descendants of the original grantees.[246]

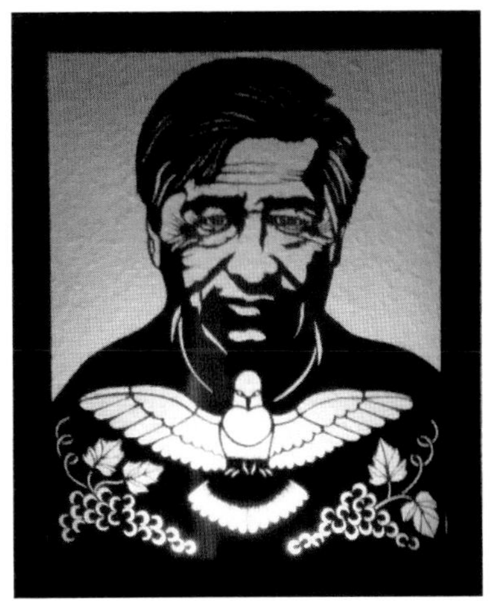

"La Paloma de Resistencia." *Papercut image of César Chávez by Valerie Rangel.*

In the 1980s, farming costs rose and farmers needed to purchase more items, including seeds, pesticides, fertilizers, machinery and fuel for the machinery, as well as pay labor wages. Supermarkets kept the purchase price of food from farmers the same, even though costs to the farmer continued to rise, which resulted in a loss of revenue.[247] When Congress passed the Immigration Reform and Control Act of 1986, the result was the first major revision of America's immigration laws in decades. The law sought to keep jobs for only those with American citizenship and aliens who were authorized to work in the United States. After years of competition with undocumented immigrant farmworkers, a cheaper labor force, American farmworkers shifted to jobs in nonagricultural sectors.[248]

By the time of César Chávez's death in 1993, the United Farm Workers union membership had fallen to less than 10 percent of its seventy-thousand-member peak of the early 1970s. During the economic recession of the 1990s, many of the Mexican workers whose immigration status had been legalized under IRCA had brought their families to the United States. The result of this influx of migrant population increased public costs for schools, health care and other services. The children of both legal and unauthorized farmworkers who had been educated in the United States did not seek labor jobs the fields as their parents once did, therefore, the cycle of dependence on fresh immigrants for cheap labor ensued.[249]

In 1994, the NAFTA trade agreement between Mexico and the United States drove down the cost of corn and coffee, forcing Mexican rural farmers into a state of bankruptcy. Extreme poverty and loss of jobs in Mexico forced rural communities to seek sources of labor, even at low wages, in the United States. By denying immigrants legal status, employers aren't obligated to pay minimum wage requirements as defined by the minimum wage law. Barack Obama advocated for the minimum wage to be raised to $9 an hour. In Florida, for example, on average, tomato farmworkers are paid based on how many buckets are picked, which comes out to about $40 a day; an average annual income is between $10,000 and $12,000.[250]

Health and Environmental Hazards

Humans have defined "pests" as plants or organisms that endanger their food supply, health or comfort. The emergence of pesticide use began in ancient times with Romans killing insects by burning sulfur and controlling weeds with salt. In the 1600s, ants were controlled with mixtures of honey and arsenic. By the late nineteenth century, U.S. farmers were using copper acetoarsenite, calcium arsenate, nicotine sulfate and sulfur to control pests in field crops, often with unsatisfactory results due to the primitive chemistry and methods of application.[251]

After World War II, chemicals such as DDT, BHC, aldrin, dieldrin and endrin were manufactured to combat weeds in grass or corn crops and to ward off insect-borne diseases such as yellow fever, encephalitis, plague, typhoid fever, malaria, dog heartworms and Rocky Mountain spotted fever.[252] Pesticides continue to be utilized worldwide. In parts of the world where some of the most volatile and hazardous chemicals have been banned, some farmers choose to continue to use pesticides despite warnings. Exposure to pesticides by farmworkers is known to cause health issues and is also thought to have adverse effects on the people who consume the contaminated products.

In a study, "Effect of Chronic Pesticide Exposure in Farm Workers of a Mexico Community," a transversal comparative study was performed between two groups, one composed of twenty-five farmworkers engaged in pesticide spraying, and a control group of twenty-one workers not exposed to pesticides; both groups belonged to the Nextipac community of Jalisco, Mexico. According to the study,

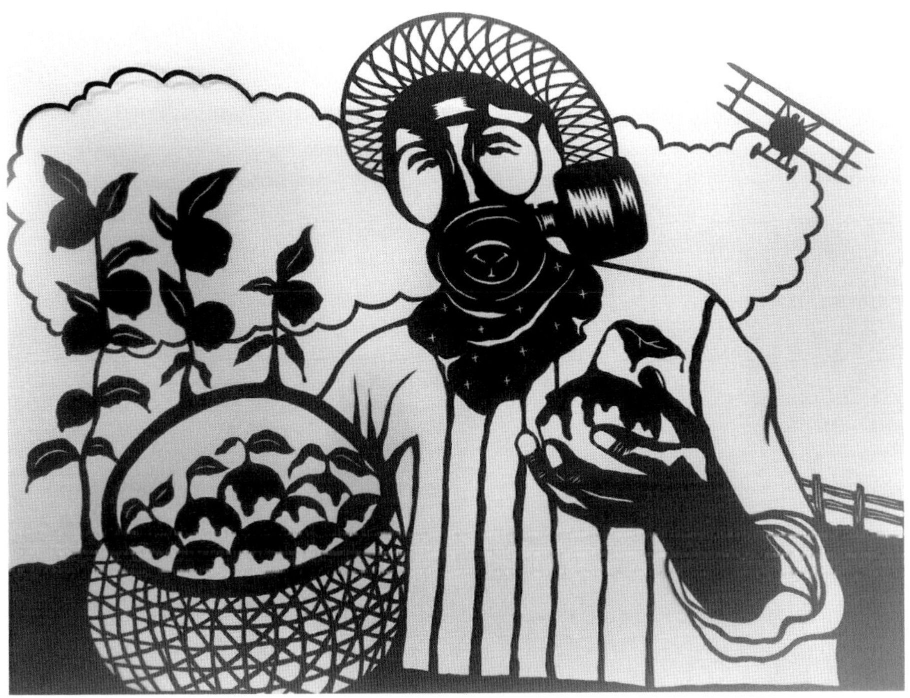

Papercut image of Pesticide Drift. *Image by Valerie Rangel.*

> *The most commonly used pesticides were organophosphates, triazines and organochlorine compounds. The exposed group, showed acute poisoning (20% of the cases) and diverse alterations of the digestive, neurological, respiratory, circulatory, dermatological, renal, and reproductive system probably associated to pesticide exposure. More importantly, they presented free DNA fragments in plasma (90.8 vs 49.05 ng/mL) as well as a higher level of lipid peroxidation (41.85 vs. 31.91 nmol/mL) in comparison with those data from unexposed farm workers. These results suggest that health hazards exist for farm workers exposed to pesticide.*[253]

Today, workers in the United States have the right to be given material safety data fact sheets for each chemical they are potentially exposed to or physically handle in their daily work operations. While not sprayed on directly, farmworkers are exposed to pesticide through the air, and many have reportedly been covered by fog and mists of pesticide in air drifts while they worked in fields. Another pathway of exposure for workers to harmful chemicals is via absorption through the skin during planting, weeding, thinning, irrigating, pruning, harvesting or while processing crops. Due to the lack of funds for out-of-pocket costs for health care and the fear of retribution from their employer, many undocumented workers do not seek the medical care they need after being exposed to pesticides, and they rarely report poisonings.[254]

Commercial farms often ignore warnings and regulations of chemicals that they use because their largely immigrant workforce does not know which chemicals are on the banned list. Also, many immigrant workers are too afraid to speak up or report violations by their employers. Many employers have reportedly threatened to "turn over" workers to immigration officials if they "speak out" and have threatened workers with the loss of jobs if they refuse to work in direct contact with pesticides. Workers are often not warned about the chemicals they are working with and do not know acceptable levels or proper handling, do not possess personal protective gear or know the symptoms of poisoning.[255]

The families of workers are also affected by pesticide and chemical application. If a worker goes home and hugs another family member, friend or loved one, they effectively spread contaminants in their clothing. If a worker brings home food that has been sprayed with chemicals or mixes laundry that has been saturated with chemicals, the entire family may have adverse reactions. Occupational and environmental health director of Farmworker Justice, Virginia Ruiz simplifies the hazards of pesticide use: "The close proximity of agricultural fields to residential areas and schools makes it nearly impossible for farmworkers and their families to escape exposure because pesticides are in the air they breathe and the food they eat, and the soil where they work and play."[256]

Long-term use of pesticides and genetically modified crops have created "super weeds" and "super pests," which have adapted to chemicals and become harder to combat. The result has been the increase of chemicals, their quantity and the application of multiple pesticides at exponential rates. Farmers are now dependent on chemicals, specialized seeds and fertilizers to maximize their crop and obtain an adequate harvest despite the potential threat that their efforts might be destroyed by super weeds and super pests resistant to chemical pesticides.

NEW MEXICO FARMWORKERS

The New Mexico Center on Law and Poverty released a report that cited an article, "New Mexico Farm Workers Forced to Work Dangerous Jobs for Little Pay." The report stated that an overwhelming majority of New Mexico farm- and dairy workers have been underpaid and mistreated, with 8 percent of them subject to wage theft and poor working conditions and forced to work overtime without lunch breaks. The report found that over 20 percent of farmworkers were exposed to pesticides while they were in the field, a violation of the Pesticide Control Act. Many experienced adverse health effects such as suffering from dizziness, lifelong respiratory issues, as well as acute poisoning. These workers could not afford health care and were untreated due to a lack of health insurance.[257]

New Mexico farmworkers are among the poorest of the working class, generating an annual family income of between $15,000 and $17,499. Of the agricultural workers surveyed in the state, the annual household income reported was $8,978. The report also stated, "New Mexican laws permit a wide swath of provisions for small family farms to keep afloat financially, but such laws are exploited by large agribusinesses as well. These provisions let farmers pay dairy workers less than minimum wage. Under these laws, agricultural workers in the state are denied other basic labor protections such as overtime pay, the ability to participate in collective bargaining, and the right to a safe and clean workplace."[258]

Why would immigrants want to come to New Mexico to work for such low wages and hazardous handling of chemicals or risk exposure to chemicals sprayed in the field? It is important to note that migrants coming to the United States may be fleeing political persecution, moving away from violence and population growth in their country of origin, or may have reduced food and employment opportunities that created economic hardships. Perhaps some migrant workers are in search of educational opportunities and adequate income to support their family back home.[259]

15

KEEP CHACO SACRED

Situated between the Ah-shi-sle-pah Wilderness Study Area (WSA) and Bisti/De-na-zin Wilderness, Chaco lands lie at the edge of oil and gas development on state and Bureau of Land Management (BLM) lands bordering the park. Archaeologists assert that Chaco Canyon was a center of trade in the Four Corners area, and culturally significant artifacts have been discovered at Chaco Canyon, which shows that this was an advanced, scientific and religious society.[260]

Despite Chaco's designation as a World Heritage Site and status as a National Historical Park, this cultural area is being infringed upon by the oil and gas industry, threatening the distinctive history and historical artifacts associated with the area. In recent years, the extractive industry has found a more cost-efficient way to expand development in the Southwest using the process of hydraulic fracking. Environmental advocates challenged the lease sale of lands near Chaco, calling for the agency to review and halt the sale.

In a letter to the Bureau of Land Management, New Mexico State Director, dated October 27, 2014, the Western Environmental Law Center, WildEarth Guardians, San Juan Citizens Alliance and the Chaco Alliance demanded that the agency stop rubber-stamping hundreds of new drilling permits in Mancos Shale/Gallup Formation wells in the San Juan basin of northwestern New Mexico. Among other concerns, the groups stated, "BLM's approvals are putting the region's cultural heritage at risk, endangering significant landscapes like Chaco Culture National Historical

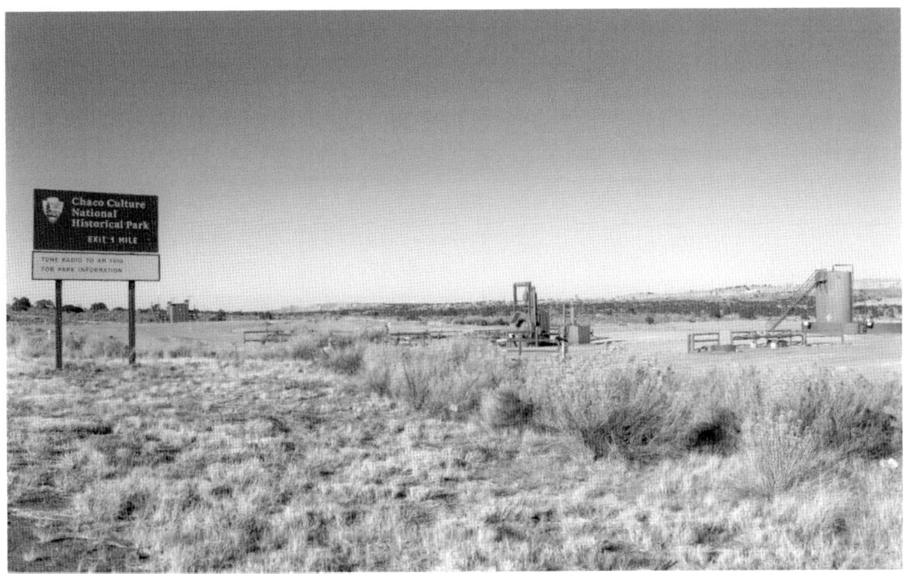

Chaco Culture National Historical Park, near fracking-threatened Chaco Canyon. *Photo credit: EvalynBemisPhotography.com.*

Park and its outliers, and other areas critical for preserving, understanding, and promoting indigenous presence in the region."[261]

On December 30, 2014, the BLM announced a decision to suspend development on public lands in northwestern New Mexico until the agency could safeguard the public health of local communities, climate, clean air and water and the region's unique cultural heritage. Kyle Tisdel, attorney and climate and energy program director for the Western Environmental Law Center, said, "In an area already besieged by oil, gas, and coal extraction, BLM cannot continue surrendering our public lands to industry without understanding the long-term cumulative effect that such development has on our health, environment, and sacred cultural resources."

Areas around Chaco Canyon have a formal wilderness designation, which serves to protect wilderness areas and prevent further oil, gas and other developments in the area. Power lines, new roadways, pipelines, pump jacks, tanks, generators and processing plants threaten to fragment wildlife habitat and impact the cultural landscape of Chaco Canyon. Flaring of oil and gas wells would result in air pollution and impact the dark night skies of the region.

On March 15, 2016, Senators Martin Heinrich and Tom Udall introduced a bill "to authorize the Secretary of the Interior to retire coal

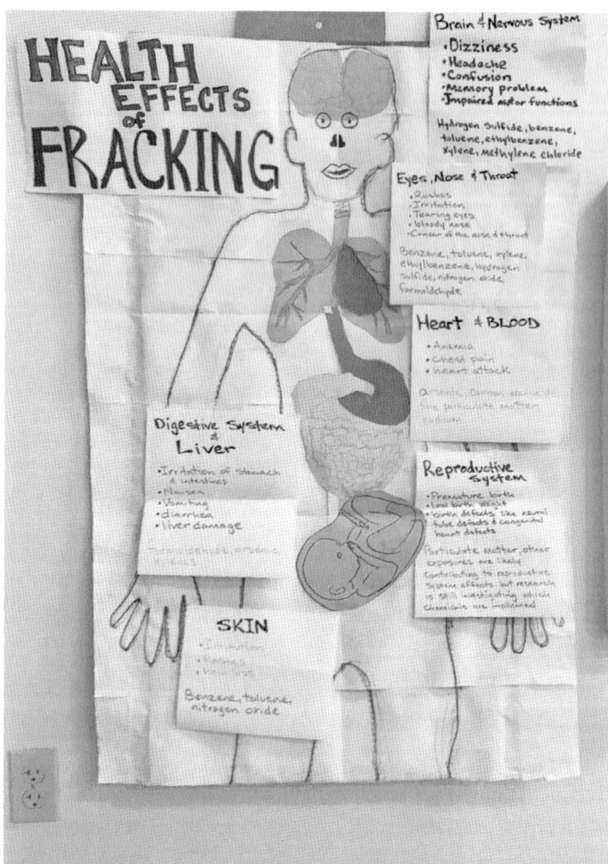

Left: Nageezi Navajo Chapter House public health poster. *Photo credit: EvalynBemisPhotography.com*.

Below: Unfenced site in Counselor, New Mexico, with horses feeding nearby. *Photo credit: Miya King-Flaherty*.

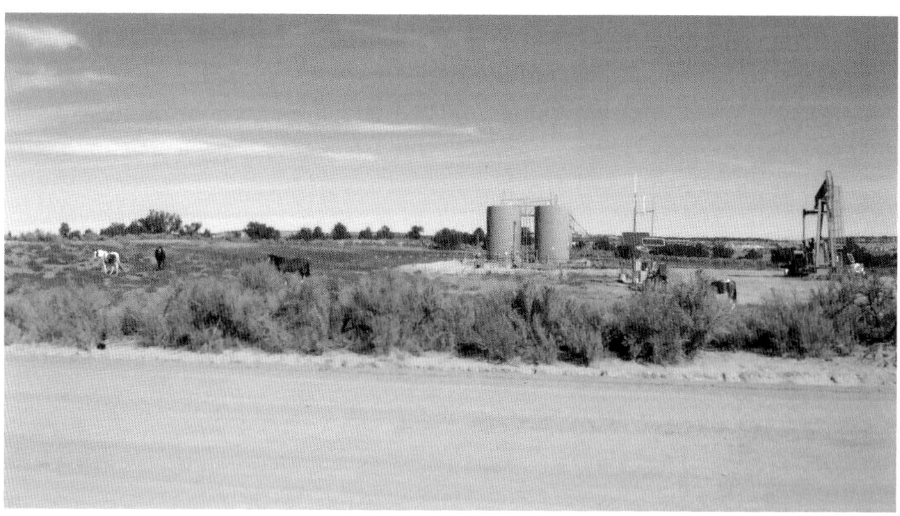

preference right lease applications for which the Secretary has made an affirmative commercial quantities determination, to substitute certain land selections of the Navajo Nation to designate as certain wilderness areas." Senator Udall's webpage shared a public "Notice of Intent" in preparation of the RMP Amendment, and an Environmental Impact Statement (EIS) was published and submitted to the Federal Register on October 21, 2016, which opened a sixty-day public scoping period that ended on December 20, 2016. The information gathered during the scoping process will be added to the information already gathered as part of BLM's prior scoping process for the EIS.[262]

The latest piece of legislation would effectively designate 7,242 acres, currently within the Ah-shi-sle-pah, as wilderness and add 2,250 acres of eligible lands to the existing Bisti/De-na-zin Wilderness, with approximately 20,000 acres of potential wilderness identified within Chaco Culture National Historical Park. The New Mexico Wilderness Alliance is also working toward legislation that would transfer adjacent state lands into park lands. To accomplish this, the state land office would have to trade certain BLM lands for the state lands, which would then be transferred to Chaco Park.[263]

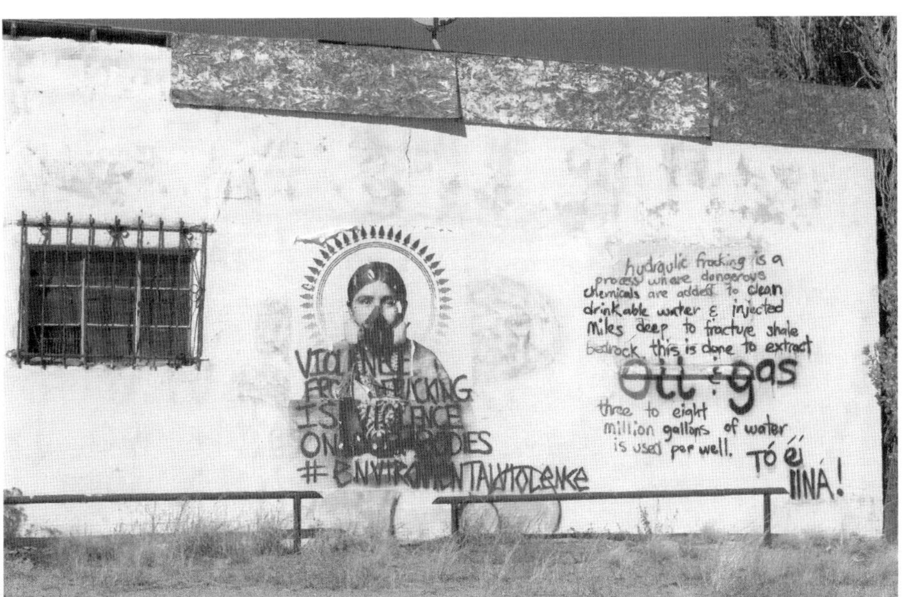

Anti-fracking graffiti art in Counselor, New Mexico. *Photo credit: EvalynBemisPhotography.com.*

As noted in an *Albuquerque Journal* article dated March 1, 2018, Interior Secretary Ryan Zinke canceled an oil and gas lease sale near Chaco Canyon. The Bureau of Land Management will hold off on the sale of twenty-five parcels on 4,434 acres within Rio Arriba, Sandoval and San Juan Counties located in northwestern New Mexico until the Department of the Interior can further review the impact on cultural artifacts in the area. In January 2018, the All Pueblo Council of Governors, representing twenty Native American tribes, formally protested the March 8 lease sale. Senator Udall responded to Zinke's news, saying, "I appreciate Secretary Zinke listening to the concerns of New Mexicans and terminating the proposed lease sale in the Chaco region." He continued, "This will provide an opportunity for the joint BLM-BIA (Bureau of Indian Affairs) regional planning process to incorporate the input of local communities, pueblos and tribes, along with industry and other stakeholders. New Mexicans deserve a say in any proposed development on public lands in our state, especially when it is near sacred or culturally sensitive land."[264]

The fate of Chaco Canyon remains uncertain. The area is one of the state's most visited tourist spots, an archeological treasure, with many New Mexicans in support of the preservation of such significant cultural resources. In an area already experiencing contamination from the oil, gas and uranium industry, further oil and gas production would increase risk to the health of the environment and surrounding communities. I hope New Mexicans "Keep Chaco Sacred."

CONCLUSION

The past is real and present, held in our memories and in the shape of the world. It is the ground of our being, its actuality, its particular substance. The future? The future doesn't exist. We must create the future by our decisions, our actions, and inactions. Together with the place we live, we are co-creators of the world, bringing it into existence moment by moment.
—Dr. Viola Cordova[265]

Our decisions, our actions, form the future and shape the health and happiness of subsequent human beings as well as the integrity and welfare of all creatures inhabiting planet earth. In a world of interconnectedness, how can we form and influence the future if we consider what was once held "sacred" as commonplace? Song, dance, language, ritual, sacred places and oral stories once inspired, captivated our imagination and imbued our spirit. Mountaintops and odd geological formations where deities held their mighty fortresses, and the places where special plants were gathered for medicine, were once so sacred that only spiritual leaders were allowed to venture, view, touch and ascend them. These sacred peaks are seen today as enterprises for capital gain, personal recreational use, or just another mound of dirt to be obliterated by dynamite and machinery in order to satiate present energy demands.

LOOK TO THE PAST

The impact of European contact on the indigenous populations of the Americas resulted in massive loss of life through infectious diseases and violent encounters that some have called the "American Indian Holocaust."[266] In the region of New Mexico, southwestern Indians encountered new diseases from colonists who displaced them from their homes and settlements. Many Indians were raped and violently attacked, their caches of food ravaged, their clothing and blankets robbed and their women and children stolen and used as slaves by Spanish settlers.

Through scientific experimentation and research with nuclear energy, the United States was first to successfully build an atomic bomb. Although scientists knew that uranium had harmful effects on human health, the bodies of indigenous peoples were disregarded, and uranium mine workers were never told about the risks associated with their work. No amount of traditional prayers or natural medicine can cure the sick and dying from uranium exposure. How do we shape our future and reconcile a past that is filled with historical trauma and actions of assault, discrimination, violence and deceit?

For the many generations that preceded colonization, indigenous communities relied on traditional cultural knowledge, natural medicine, prayers and traditions passed down through oral history. An unintentional result of Western colonial expansion was the decline of indigenous populations; the loss of language, traditions and knowledge has resulted in a transgenerational fear of Western medicine and education. Understanding the social and environmental history of man provides a knowledge base for what has occurred so that we can envision a better future.

ASSESS THE PRESENT DANGER

The following quote from Gerald Vizenor offers an indigenous perspective on colonialism:

> *A focus on community resilience moves beyond personal traits and abilities to emphasize systemic and structural issues that might be causes of or solutions to personal and community suffering. Examining community resilience is helpful in both recognizing the devastating effects of colonialism*

> *and genocide as well as recognizing the persistence and survival that American Indian communities embody today.*[267]

Historical trauma is a term that describes the impacts of cultural suppression, oppressive policies and traumatic experiences of indigenous peoples in North America as a result of colonization.[268] Practices of leasing Indian lands for militarization and experimentation, mineral and resource extraction, pumping groundwater and dumping pollutants, exploitation of cheap migrant farmer labor, energy extraction and power generation, as well as industrial development have scarred the land, leaving irreversible environmental degradation that has affected the health of the surrounding communities.

As a way to heal from historical trauma and the issues associated with marginalization, indigenous language and religious suppression, forced assimilation and residential/boarding schools, which changed the culture of Native Americans and communities of color, there must first be an acknowledgement of the traumatic history that has taken place. Government and industries must recognize the process of judgement and decision making that led to policies that allowed for the destruction of the environment and dumping on poor, marginalized communities of color. Emphasis should be made on revitalizing culture of indigenous communities through the perpetuation of traditional ecological knowledge and the restoration of native languages, stories and rituals. Within these sources of knowledge lie moral codes of environmental and community stewardship, and the belief of interconnectedness.

In the twentieth century, bureaucratic forms of oppression continue, such as the portrayal of Native Americans as mascots or savages, uncivilized, incapable of managing land, resources and animals, unable to educate their own children or make healthy decisions for themselves. Mainstream television, media ads and news channels do not feature the strife and struggles of marginalized communities of color. Therefore, respect and value of diverse communities of color must resonate through mainstream culture.

THE FUTURE

Based on research, there is a causal relationship between heat and violence, therefore, increasing global temperatures will likely result in an increase in aggression.[269] Diminishing resources are likely to increase conflict between groups or the migration of one group to another group's territory, leading to conflict over rights and ownership of space.[270] If governments fail to adequately protect against natural disasters or respond to their effects, people lose confidence and trust in authority.[271]

Nations and communities possessing fewer resources are likely to feel the impacts of climate change more acutely, as they have less ability to afford the technologies and medications to mitigate effects of change. Within nations, individuals of lower socioeconomic status are more likely to become ethnic minorities, increasing ethnic tensions and hostility.[272] Forced relocations and land displacements in New Mexico have resulted in the loss of physical, geographic and social connections within communities, which has resulted in grief, anxiety, fear, wariness of strangers and political entities, as well as a sense of loss of language, family and cultural identity.

What does a healthy community and ecosystem look like, and is it even possible? As part of a community health assessment, the history and land use of a community, social determinants of health, environmental factors and issues of historical trauma are all considered. Communities experiencing environmental injustice experience health disparities that are often the result of economic inequalities, lack of education, social and genetic risk factors—all of which ultimately cause human suffering.

Actions that shape the future must include advocacy for systemic change. Structural change at the government-to-government level in the cases of American Indian and Alaskan natives starts with honoring treaty rights and federal obligations to provide for the health and well-being of these citizens. State and federal agencies must ensure compliance with EPA standards and stricter standards set by sovereign Indian nations. New technology and the pursuits of industry, as well as the reopening of mining operations in communities already affected by contamination are decisions that must be made with consideration for social equity and environmental justice.

Care and consideration of the cumulative effects from historical environmental contamination must also be weighed when dealing with toxic sources of contamination, and preventative measures must be taken to avoid a disproportionate number of health hazards affecting poor, marginalized communities of color. Restoration of the environment begins with the initiation of cleanup practices, ongoing monitoring, community awareness building, involvement of volunteers within communities and environmental stewardship. Adequate compensation must come to communities whose health has been impacted by government, private industry, illegal dumping and militarization.

Communities must be the collective force of change by defining and advocating for social and environmental justice; utilizing open public community discussion forums to allow individuals of adversely affected communities the opportunity to voice their concerns on issues. Talking circles and discussion forums validate each person's struggles, allowing information to be shared and individuals to network, problem-solve and build support systems. While long-term goals of environmental remediation, social economic justice and health equity are possible to achieve, it is also important to focus on short-term goals such as individual mental and physical wellness, as well as strengthening the diverse cultural traditions of New Mexican communities.

The dream of healing past wrongs and preventing future generations of communities of color from bearing unequal burdens of waste by sources of contamination starts with the actions of today. Environmental injustice harms not only the communities it affects; it also harms man's social structure, causing mistrust, war and civil unrest—this then affects all future land use and planning of our precious planet's natural resources. Man's ingenuity must stem from moral codes of conduct imbedded within the tales of creation, with reverence for the earth's health and sustainability its leading principles.

EPILOGUE

In 2021, the City of Santa Fe and County proposed the San Juan-Chama Return Flow Project to increase the long-term security and climate resilience of its municipal water supply. The project includes a seventeen-mile pipeline that will return treated, unconsumed San Juan-Chama water from the Paseo Real Water Reclamation Facility back to the Rio Grande, allowing the City to release less water from upstream reservoirs on the Rio Grande while diverting the same amount of water at the Buckman Direct Diversion.[273]

In March 2022, Senator Tom Udall introduced a bill in Congress for the third time designating segments of the Gila River as part of the National Wild and Scenic Rivers System for protection. The bill would mandate consultation with tribes for a comprehensive management plan and create a native fish habitat restoration project for the recovery of a threatened or endangered species. A diverse coalition hopes to prevent future dams from being built on the Gila and San Francisco Rivers. Lori Gooday Ware, chairwoman of the Fort Sill Apache Tribe, said the bill is needed to ensure traditional and current use of the waterways, and protect critical wildlife habitat. She wants her grandchildren to experience the rivers the way their indigenous ancestors once did.[274]

During the summer of 2022, the Black Fire burned 325,136 acres of the Gila National Forest. Heavy rains in August caused the Gila River to crest at 30.28 feet, breaking the all-time record water level for that gauge site near Virden, New Mexico.[275] In February 2023, the U.S. Forest Service was authorized to shoot wild cattle that were causing environmental damage by compromising water quality and habitat for other species by trampling stream banks in sensitive areas in the Gila Wilderness. Attorneys for ranchers in opposition claimed shooting as many as 150 "unauthorized" cows on public land was a violation of federal regulations and amounted to animal cruelty. The cattle were descendants of cows that legally grazed the area in the 1970s before the ranch went out of business. Unauthorized livestock, as defined in Forest Service regulations, refers to any cattle, sheep, goats or hogs that are not authorized by permit to be grazing on national forest land. Regulations call for an impoundment order to be issued and the livestock rounded up, with lethal action being a final step for those that aren't captured. Due to the rugged terrain of the Gila National Forest, cattle could not be rounded up and evacuated, therefore the Forest Service called for shooting the cattle with a high-powered rifle from a helicopter and leaving the carcasses in the Gila Wilderness; an estimated sixty-five tons of dead animals would be left in the forest for months until they decompose or are eaten by scavengers, and lead to contamination of waterways.[276]

In January 2021, a ten-year plan was developed in cooperation with multiple federal partner agencies including Bureau of Indian Affairs, Department of Energy, Nuclear Regulatory Commission, Navajo Area Indian Health Service and the Agency for Toxic Substances and Disease Registry to incorporate goals for achieving assessment and cleanup actions for uranium contamination on the Navajo Nation.[277] The EPA and NNEPA prioritized forty-six mines based on gamma radiation levels, proximity to homes and potential for water contamination identified in preliminary assessments. Forty-four mines are in the assessment phase, which includes biological and cultural surveys, radiation scanning and soil and water sampling.[278] In 2022, New Mexico lawmakers passed a bill to develop a strategic plan for uranium cleanup and to focus economic development on reclamation. Cleanup of hundreds of abandoned mines will finally begin once financial settlements are reached with large corporations and the U.S. government for decades of contamination.

In March 2022, the U.S. Army reached an agreement with Zuni Pueblo and the Navajo Nation and the state of New Mexico to pay $1.5 million toward restoring environmental damage done at a former munitions depot. The proposed settlement allots $1 million for restoration projects in addition to ongoing cleanup at the site, $117,000 for cultural services damage and $314,000 to cover past and future costs of the state Natural Resources Trustee's Office. The cleanup work also involves finding, disarming and removing explosives.[279]

A $32-million settlement over the 2015 Gold King Mine was reached in 2022. The agreement mandates the federal government to make cash payments for response costs, environmental restoration, water quality monitoring and cleanup activities, as well as help mitigate negative perceptions about the area's rivers following the spill. The state received $11 million in damages from the mining companies, and Colorado and the Navajo Nation also made multimillion-dollar agreements to settle claims for cleanup of the Superfund site that was established following the spill; the case against the federal contractors involved is pending.[280]

During the coronavirus pandemic, imposed lockdown orders, inadequate infrastructure and lack of access to basic needs like running water led to the highest coronavirus infection and mortality rates in the country among Navajo People.[281] In June 2023, groundwater levels for the City of Gallup and surrounding Navajo communities dropped two hundred feet, resulting in over 40 percent of Navajo households relying on hauling water to meet their needs. After San Juan Generating Station's closure in June 2022, PNM began working with the U.S. Bureau of Reclamation, the Navajo Nation and the City of Gallup to transfer the station's water infrastructure over to the Navajo-Gallup Water Supply Project to help Navajo communities obtain drinking water.[282]

In 2021, DOE submitted a permit renewal application to NMED that removed WIPP's 2024 closure date, leaving its lifetime open-ended. Then in October 2022, DOE proposed a dramatic expansion of the type and amount of radioactive waste for burial at WIPP and called on LANL to produce thirty plutonium pits annually for nuclear armament by 2026. The Savannah River Site in South Carolina was also tasked with producing fifty pits annually by 2030. Plans call for shipping material to LANL, where it would be converted to oxidized powder, then transported to South Carolina to manufacture weapons with excess waste sent to WIPP.[283]

LANL is located within the Jemez Mountains, which are volcanic mountains that erupted intermittently 14 million years ago to as recently as 40,000 years ago with a supervolcano, the Valles Caldera, one of the largest young calderas on earth in its midst.[284] Geologists noted that the presence of hot springs shows that the caldera is part of a large, long-lived and *still active* system. According to the New Mexico Bureau of Geology and Mineral Resources' 2012 publication, both local and global seismic tomography and recent GPS data suggests that mantle is passively upwelling into the thinned, stretched lithosphere in the Rio Grande Valley, rifting it apart. The widening is not limited to the rift, but also affects the western Great Plains and the eastern Colorado Plateau, where Los Alamos is located.[285]

Oil companies operating in the most active oilfield in the United States, the Permian Basin, also oppose the plan to store spent nuclear fuel from commercial power plants at the proposed Holtec facility near Hobbs, New Mexico. In 2022, sixty-three earthquakes were recorded near Hobbs: thirty-five earthquakes were between 3.0 and 4.0 magnitude, ninteen were between 4.1 and 6.0 magnitude.[286] A final environmental impact statement on the Holtec project was released in July 2022 by the federal Nuclear Regulatory Commission (NRC) with a finding of minimal impact to New Mexico for the proposed facility, which would have capacity to store 173,600 metric tons of high-level nuclear waste from commercial power plants all over the country. Shipments of nuclear waste would travel by rail through many parts of the state.[287] State Attorney General Hector Balderas filed a lawsuit against the NRC disputing its environmental analysis as inadequate. The case was dismissed by U.S. District Judge James O. Browning, stating that the court did not have jurisdiction to rule in the matter.[288]

In January 2023, lawmakers introduced a bill that would mend New Mexico's Radioactive and Hazardous Materials Act, prohibiting state agencies from issuing permits to certain types of waste storage facilities.[289] The Western Governors' Association called on the federal government to devise regulations that include state consent when siting and licensing facilities to store nuclear waste, and also demanded that no temporary storage sites be developed until a permanent repository is available to take the waste; states must also be consulted on transportation routes, operations and preparation in case of any associated accidents or radiation releases.[290]

On June 2, 2023, Secretary of the Interior Deb Haaland issued a public land order withdrawing public lands within a ten-mile radius of the Chaco Culture National Historical Park from new oil and gas leasing and mining claims for twenty years to better protect the sacred and historic sites and tribal communities currently living in northwest New Mexico. The withdrawal applies only to public lands and federal mineral estate and does not apply to minerals owned by private, state or tribal entities. It does not affect valid existing leases; during the twenty-year withdrawal period, production from existing wells could continue, additional wells could be drilled on existing leases and Navajo nation allottees can continue to lease their minerals.[291]

NOTES

Introduction

1. World Health Organization, accessed February 1–April 30, 2017, http://www.who.int/globalchange/ecosystems/en.
2. Ibid.
3. Millennium Ecosystem Assessment, accessed February 1–April 30, 2017, http://www.millenniumassessment.org/en/About.html.
4. Edward O. Wilson, *Consilience: The Unity of Knowledge*, accessed March 9, 2018, http://www.wtf.tw/ref/wilson.pdf.

Chapter 1

5. New Mexico Secretary of State: Native Americans, accessed January 4, 2017, www.sos.state.nm.us.
6. New Mexico Office of the State Historian: Ancient Peoples of New Mexico, accessed January 4, 2017, newmexicohistory.org/people/ancient-peoples-of-new-mexico.
7. Joe S. Sando, *Pueblo Nations: Eight Centuries of Pueblo Indian History* (Santa Fe, NM: Clear Light Publishers, 1992), 28–53.
8. Ibid.
9. Bittanie Smith, "Creating Numbe Whageh: An Uphill Battle between Cultures," *University Daily Kansan*, accessed February 10, 2017, http://www.

kansan.com/arts_and_culture/creating-numbe-whageh-an-uphill-battle-between-cultures/article_246b2b92-efb2-11e6-883c-97d4929f481a.html.

10. Leonard Peltier, statement at extradition hearing, Vancouver, British Columbia, May 13, 1976.

11. Donald A. Grinde, Bruce E. Johansen and Howard Zinn, *Ecocide of Native America: Environmental Destruction of Indian Lands and Peoples* (Santa Fe, NM: Clear Light Publishers, 1995).

12. Ibid.

13. Civil War Muster rolls, New Mexico State Archives.

14. The Bosque Redondo Memorial, accessed February 1–April 30, 2017, https://www.bosqueredondomemorial.com/manifest_destiny.htm.

15. Oral story of a Navajo weaver, Roseann Willink, 2005.

16. The Bosque Redondo Memorial, accessed February 1–April 30, 2017, https://www.bosqueredondomemorial.com/manifest_destiny.htm.

17. Susan Perlman, 1997. Fort Wingate Depot Activity Ethnographic Study. Albuquerque, New Mexico: Office of Contract Archeology. University of New Mexico.

18. Neal W. Ackerly, *A Navajo Diaspora: The Long Walk to HWÉELDI* (Silver City, NM: Dos Rios Consultants, 1998).

19. Laurance Linford, *Navajo Places: History, Legend, Landscape: A Narrative of Important Places On and Near the Navajo Reservation, with Notes on Their Significance to Navajo Culture and History* (Salt Lake City: University of Utah Press, 2000).

20. Fort Wingate, GlobalSecurity.org, 2000, accessed January 1–April 30, 2006, http://www.globalsecurity.org/military/facility/fort-wingate.htm.

21. Department of Defense Environmental Cleanup. 2005. Military History of Fort Wingate, accessed October 10, 2005, http://www.dtic.mil/envirodod/derpreport/wingate.html.

22. Plan for the Use of Fort Wingate, prepared by the New Mexico Department of Game and Fish, 1995.

23. Schelberg, John. *Repatriation of Fort Wingate Army Depot Activity Area.* U.S. Army Corps of Engineers, Albuquerque District.

24. Plan for the Use of Fort Wingate.

25. "Church Rock Tailings Spill," The Energy Library, accessed January 1–April 30, 2006, http://www.theenergylibrary.com.

26. 96th Congress, House Committee on Interior and Insular Affairs, Subcommittee on Energy and the Environment, "Mill Tailings Dam Break at Church Rock, New Mexico" (hereafter cited as "Church Rock Tailings Spill"), October 22, 1979.

27. Jorge Winterer, "Potential Health Impact of United Nuclear Church Rock Spill," *American Medical Association Journal of Ethics* (Fall 1979); Gallup: Physicians for Social Responsibility.
28. Kathie Saltzstein, "Navajos Ask $12.5 Million in UNC Suits," *Gallup Independent*, August 14, 1980.
29. "Church Rock Tailings Spill."
30. Saltzstein, "Navajos Ask $12.5 Million in UNC Suits."
31. "Church Rock Tailings Spill."
32. Ibid.
33. Marjorie Childress, "Church Rock Uranium Mining Can't Start Just Yet. EPA Says the Land Is Subject to Regulation under the Safe Drinking Water Act," *New Mexico Independent*, March 18, 2010.
34. Southwest Research and Information Center, "Navajo Nation President Joe Shirley, Jr. Signs Diné Natural Resources Protection Act of 2005; New Law Bans Uranium Mining, Processing Throughout Navajo Nation," accessed April 30, 2005, http://www.sric.org/voices/2005/v6n2/navajo_pr_dnrpa.php.
35. Saltzstein, "Navajos Ask $12.5 Million in UNC Suits."
36. Jorge Winterer, "Potential Health Impact of United Nuclear Church Rock Spill," Physicians for Social Responsibility, Gallup, New Mexico, Fall 1979.
37. Hanksville Organization. "KILLING OUR OWN, The Disaster of America's Experience with Atomic Radiation," *Part 10 of 19: Uranium mining and the Church Rock Disaster*, accessed January 1–April 30, 2006, http://www.hanksville.org/voyage/misc/uranium/ChurchRock.html.
38. Ibid.
39. "Church Rock Spill," *Albuquerque Journal*, July 17, 1980.
40. Southwest Research and Information Center, "Navajo Nation President Joe Shirley, Jr. Signs Diné Natural Resources Protection Act."
41. "Church Rock Spill."

Chapter 2

42. Grinde, Johansen and Zinn, *Ecocide of Native America*.
43. "Mining Artifacts," *New Mexico Mines*, accessed November 19, 2016, http://www.miningartifacts.org/New-Mexico-Mines.html.
44. New Mexico Art Museum, "New Mexico Art."
45. Ibid.

46. Ibid.
47. "Church Rock Tailings Spill."
48. Ibid.
49. Christopher McLeod, Randy Hayes and Glenn Switkes, foreword by David R. Brower, *The Four Corners: A National Sacrifice Area Resource Guide*, video recording with twenty-four-page transcript (Oley, PA: Bullfrog Films, University of California, Berkeley, Graduate School of Journalism, Earth Island Institute, 1985).
50. Anne Minard, "Diné CARE, Environmental Groups Sue Interior Over Navajo Coal Plant," *Indian Country Today*, accessed May 5, 2016, http://indiancountrytodaymedianetwork.com/2016/05/05/dine-care-environmental-groups-sue-interior-over-navajo-coal-mine-164362.
51. Navajo Nation Human Rights Commission, *The Impact of the Navajo-Hopi Land Settlement Act of 1974 P.L. 93-531 et al*, Public Hearing Report, 2012, accessed April 1–May 31, 2017, http://www.nnhrc.navajo-nsn.gov/docs/NewsRptResolution/070612_The_Impact_of_the_Navajo-Hopi_Land_Settlement_Act_of_1974.pdf.
52. "About Coal Mining Impacts," Greenpeace.org, accessed January 2017, http://www.greenpeace.org/international/en/campaigns/climate-/change/coal/Coal-mining-impacts//.
53. L.M. Shields, W.H. Wiese, B.J. Skipper, B. Charley and L. Benally, "Navajo Birth Outcomes in the Shiprock Uranium Mining Area," *Health Physics* 63 (November 1992): 542–551, accessed April 1–May 31, 2017, http://www.sric.org/uranium/navajorirf.php.
54. Navajo Nation Human Rights Commission, *The Impact of the Navajo-Hopi Land Settlement Act*.
55. Max Goldtooth, Peter MacDonald and Ron Milford, "Navajo Water Rights: Truths and Betrayals," *High Country News*, July 21, 2008, accessed November 18, 2016, http://www.hcn.org/issues/40.13/navajo-water-rights-truths-and-betrayals.
56. Marlon Duke, "Navajo-Gallup Water Supply Project Progresses with Two New Contracts Awarded," U.S. Bureau of Reclamation, September 20, 2016, accessed November 18, 2016, http://www.usbr.gov/newsroom/newsrelease/detail.cfm?RecordID=56687.
57. Goldtooth, MacDonald and Milford, "Navajo Water Rights."
58. Ibid.
59. Grinde, Johansen and Zinn, *Ecocide of Native America*.
60. Duke, "Navajo-Gallup Water Supply Project Progresses."
61. Ibid.

62. Winona LaDuke, "Monster Slayers: Can the Navajo Nation Kick the Coal Habit?," *Indian Country Today*, July 31, 2013, accessed March 9, 2018, http://indiancountrytodaymedianetwork.com/2013/07/31/monster-slayers-can-navajo-nation-kick-coal-habit-150637.
63. McLeod, Hayes and Switkes, *The Four Corners*.

Chapter 3

64. "Documentary Focuses on Navajo Water Rights on PBS," *Arizona Daily Sun*, 2007, accessed January 11, 2017, http://azdailysun.com/news/local/documentary-focuses-on-navajo-water-rights-on-pbs/article_160062b2-a160-5394-aec3-968d79586a04.html.
65. Erica Gies, "The Navajo Are Fighting to Get Their Water Back," Takepart.com, April 22, 2016, accessed January 11, 2017, http://www.takepart.com/feature/2016/04/22/native-american-water-settlements.
66. Office of the State Engineer, "Navajo Nation Water Resource Development Strategy (Draft)," July 2011, Navajo Nation Department of Water Resources.
67. "Documentary Focuses on Navajo Water Rights," *Arizona Daily Sun*.
68. Gies, "The Navajo Are Fighting."
69. Office of the State Engineer, Interstate Stream Commission, "Water Rights in the San Juan River Basin in New Mexico," accessed January 1–May 31, 2017, http://www.ose.state.nm.us/Basins/Colorado/isc_CO_SJwaterRights.php.
70. EPA, "Navajo Nation." *Unregulated Water Source Sampling Results*, October 2009, accessed January 1–May 31, 2017, https://www.epa.gov/sites/production/files/2016-06/documents/2010-08-01-navajo-water-sample-results-final-report.pdf.
71. Office of the State Engineer, "Water Rights in the San Juan River Basin."
72. David J. Wishart, *Encyclopedia of the Great Plains*. 2011; *Winters v. United States*, 207 U.S. 56 (1908), Supreme Court case; *Winter's Doctrine*, accessed January 1–May 31, 2017, http://plainshumanities.unl.edu/encyclopedia/doc/egp.wat.041.
73. United Nations, "Human Right to Water and Sanitation," http://www.un.org/waterforlifedecade/human_right_to_water.shtml.
74. Office of the State Engineer, "Water Rights in the San Juan River Basin."
75. U.S. Department of the Interior, "Sec. Salazar Signs Decision on Navajo-Gallup Water Supply Project, Clearing Way for Historic Water Rights

Settlement," 2009, accessed March 9, 2018, https://www.doi.gov/news/pressreleases/2009_10_01_releaseA.

76. United Nations, "Human Right to Water and Sanitation."
77. Ibid.
78. Ibid.
79. Ibid.
80. U.S. Department of the Interior, "Sec. Salazar Signs Decision."
81. Ibid.
82. Ibid.
83. Ibid.
84. Ibid.
85. Ibid.
86. John Fleck, "Whitehorse Lake Sees Flowing Water at Last," *Albuquerque Journal*, accessed January 5, 2014, https://www.abqjournal.com/331261/whitehorse-lake-sees-flowing-water-at-last.html.

Chapter 4

87. National Trust for Historic Preservation, "Saving Places: Mount Taylor," accessed November 22, 2016, http://www.preservationnation.org/travel-and-sites/sites/southwest-region/mount-taylor.html.
88. "Mount Taylor's Spiritual and Cultural Values Merit New Protection," *New Mexico Independent*, accessed January 1–May 31, 2017, http://newmexicoindependent.com/18977/mount-taylors-spiritual-and-cultural-value-merit-new-protection.
89. Alysa Landry, "Tribes Fight to Regain Traditional Cultural Property Designation for Mount Taylor," *Indian Country Today*, 2012, accessed January 1–May 31, 2017, http://indiancountrytodaymedianetwork.com/2012/12/02/tribes-fight-regain-traditional-cultural-property-designation-mount-taylor-145985.
90. David Gerard, *The Mining Law of 1872: Digging a Little Deeper*, Property and Environmental Research Center, accessed January 1–May 31, 2017, http://www.perc.org/articles/mining-law-1872-0.
91. Robert Tohe, "New Uranium Mine Threatens Mount Taylor," Rio Grande Sierra Club, accessed November 22, 2016, http://www.riograndesierraclub.org/new-uranium-mine-threat-to-mount-taylor.
92. Ibid.
93. Ibid.

94. Ibid.
95. Staci Matlock, "Court Ruling Will Help Protect Mount Taylor," *Santa Fe New Mexican*, February 6, 2014, accessed November 22, 2016, http://www.santafenewmexican.com/news/local_news/court-ruling-will-help-protect-mount-taylor/article_660a0a06-da66-5fb8-b7a1-0611c10dc293.html.
96. Ibid.
97. Ibid.
98. *Los Alamos Heritage*, "Los Alamos: Where Discoveries Are Made," accessed March 9, 2018, http://www.visitlosalamos.org/heritage; http://www.visitlosalamos.org.
99. Ibid.
100. Jacqueline Devine, "Tularosa Downwinders to Protest at Trinity Site," *USA Today*, 2016, accessed November 29, 2016, http://www.usatoday.com/story/news/local/community/2016/04/01/tularosa-downwinders-protest-trinity-site/82541134.
101. Ibid.
102. Ibid.
103. "Cyanide, Other Poisons Found in Los Alamos Storm Runoff," *Albuquerque Tribune*, September 12, 2000, A2.
104. The New Mexico Environment Department, "Wildfire Impacts on Surface Water Quality," 2013, accessed January 1–May 31, 2017, https://www.env.nm.gov/swqb/Wildfire.
105. Nuclear Active (organization), "In Rush to Reopen WIPP, DOE Ignores the Safer 'Clean Salt' Option," November 18, 2016, accessed January 1–May 31, 2017, http://nuclearactive.org/in-rush-to-reopen-wipp-doe-ignores-the-safer-clean-salt-option.
106. Ibid.
107. Brett Madres, "Storage and 'Disposal' of Nuclear Waste," March 18, 2011, accessed February 1, 2018, http://large.stanford.edu/courses/2011/ph241/madres1.
108. Nuclear Waste Aqui (organization), "Citizens Oppose HR 3053—Nuclear Waste Policy Act Amendments of 2017" (press release), accessed February 1, 2018, http://nonuclearwasteaqui.org/2017/10/radioactive-waste-bill-threatens-texas-and-new-mexico-poses-nationwide-risks-from-dangerous-unnecessary-transport.
109. Rodney C. Ewing, "Long-term Storage of Nuclear Fuel," *Nature*, 2006, accessed January 1–May 31, 2017, http://www.nature.com/articles/nmat4226?WT.ec_id=NMAT-201503.

110. S.E. Hasan, *International Practice in High-Level Nuclear Waste Management in Concepts and Applications in Environmental Geochemistry*, edited by D. Sarkar, R. Datta and R. Hannigan (Boston, MA: Elsevier, 2007).
111. Nuclear Waste Aqui, "Citizens Oppose HR 3053."
112. Ibid.

Chapter 5

113. Jerry B. Howard, "Hohokam Legacy: Desert Canals." *Pueblo Grande Museum Profiles No. 12*, accessed March 16, 2012, http://www.waterhistory.org/histories/hohokam2.
114. USDA Forest Service, "Gila National Forest," 2013, accessed March 16, 2012, https://web.archive.org/web/20060111100317/. http:/www2.srs.fs.fed.us/r3/gila/about.
115. William K. Hartmann and Gayle Harrison Hartmann, "Juan de la Asunción, 1538: First Spanish Explorer of Arizona?" *Kiva* 37, no. 2 (1972): 93–103.
116. Jonathan Waterman, "American Nile: Saving the Colorado," *National Geographic*, accessed January 2, 2017, http://www.nationalgeographic.com/americannile.
117. David Roberts, *Once They Moved Like the Wind: Cochise, Geronimo, and the Apache Wars* (New York: Touchstone, 1994).
118. Charles S. Peterson, *Pioneer Settlements in Arizona* (New York: Macmillan, 1992), accessed March 16, 2012, http://www.lightplanet.com/mormons/daily/history/gathering/Arizona_EOM.htm.
119. U.S. Geological Survey, 1914–2011, National Water Information System, "USGS Gage #09432000 on the Gila River Below Blue Creek, Near Virden, NM," accessed March 16, 2012, https://wdr.water.usgs.gov/wy2011/pdfs/09432000.2011.pdf.
120. Michael J. Cohen, Christine Henges-Jeck and Gerardo Castillo-Moreno, "A Preliminary Water Balance for the Colorado River Delta, 1992–1998," *Journal of Arid Environments* 49 (2001): 35–48, accessed March 16, 2012, http://www.pacinst.org/reports/missing_water/missing_water_article_web.pdf.
121. The Nature Conservancy, "Gila Flow Needs Assessment," accessed July 2014, http://nmconservation.org/Gila/GilaFlowNeedsAssessment.pdf.
122. Chris Williams, "The Battle to Save New Mexico's Last Wild River," Truth-out (organization), accessed September 11, 2011, http://www.truth-out.org/news/item/34331-the-battle-to-save-new-mexico-s-last-wild-river.

123. Barry Massey, “NM Governor Pledges to Fight Gila River Diversion,” *Las Cruces Sun-News*, 2008, accessed June 11, 2008, https://web.archive.org/web/20080611144914/http:/www.lcsun-news.com/ci_8960289.
124. National Geographic, “Still Wild and Free, New Mexico’s Gila River Is Again Under Threat,” accessed September 11, 2001, http://voices.nationalgeographic.com/2011/09/27/still-wild-and-free-new-mexicos-gila-river-is-again-under-threat.
125. Ibid.
126. Ibid.
127. Ibid
128. Williams, “The Battle to Save New Mexico’s Last Wild River.”

Chapter 6

129. Colorado River Water User Association, “New Mexico,” accessed December 11, 2016, https://www.crwua.org/colorado-river/member-states/new-mexico.
130. Dale Rodebaugh, “The Eagle: At Home Along the Animas,” *Durango Herald*, accessed July 8, 2012, https://durangoherald.com/articles/37713.
131. Colorado River Outfitters Association, “Commercial River Use in the State of Colorado, 1988–2011,” 2012, accessed July 8, 2012, http://www.croa.org/media/documents/pdf/2011-commercial-rafting-use-report-final.pdf.
132. Colorado Parks and Wildlife, “Colorado Fishing,” 2014, accessed October 21, 2014, http://cpw.state.co.us/Documents/RulesRegs/Brochure/fishing.pdf.
133. Jen Wolf, “The River of Lost Souls,” See the Southwest, 2012, accessed December 11, 2016, http://seethesouthwest.com/4472/the-river-of-lost-souls.
134. Ibid.
135. U.S. Department of the Interior. Bureau of Reclamation. *Amended and Restated Agreement in Principle Concerning the Colorado Ute Indian Water Rights Settlement and Binding Agreement for Animas-LaPlata Cost Sharing.* https://www.usbr.gov/uc/progact/animas/pdfs/1_ALPCostSharingAgt313_02.pdf. Accessed: October 21, 2014.
136. Ibid.
137. Jonathan Thompson, “Animas River Spill: Only the Latest in 150 Years of Pollution. Mapping the Other Threats to the Animas and San Juan

Rivers," *High Country News*, accessed December 1, 2016, http://www.hcn.org/articles/beleaguered-watershed-animas-spill-epa-durango.

138. Dan Elliot, "EPA Rejects $20.4 Million in Requests for Gold King Mine Spill Costs," *Denver Post*, December 9, 2016, accessed December 11, 2016, http://www.denverpost.com/2016/12/09/epa-gold-king-mine-spill-costs. Accessed: December 11, 2016.
139. Ibid.
140. Ibid.
141. Ibid.
142. Ibid.
143. Jonathan Thompson, 2015. "When Our River Turned Orange. Nine Things You Need to Know about the Animas River Mine Waste Spill," *High Country News*, accessed December 11, 2016, http://www.hcn.org/articles/when-our-river-turned-orange-animas-river-spill.
144. United States Department of Labor, Occupational Safety and Health Administration. "OSHA: Heavy Metals Information," accessed December 11, 2016, https://www.osha.gov/SLTC/metalsheavy.
145. Navajo Nation, "Navajo Nation Victorious in Initial Gold King Mine Spill Ruling" (press release), Navajo Nation Administrative Legal Secretary, Annabelle Henderson, accessed February 14, 2018.
146. Ibid.
147. Elliot, "EPA rejects $20.4 million in requests."
148. Navajo Nation, "Navajo Nation Victorious."

Chapter 7

149. Central downtown Albuquerque history and architecture, "Albuquerque: A History," accessed January 2017, http://www.downtownacd.org/wp-content/uploads/2013/08/Central-Downtown-History-Architecture-.pdf.
150. Susan DeWitt, *Historic Albuquerque Today, An Overview Survey of Historic Buildings and Districts* (Second edition) (Albuquerque, NM: Historic Landmarks Survey of Albuquerque, 1978).
151. V.B. Price, *The Orphaned Land: New Mexico's Environment Since the Manhattan Project* (Albuquerque: New Mexico Press, 2011).
152. Hugh G. Bennett, testimony in Congress, House, Subcommittee of the Committee on the Public Lands, Soil Erosion Program, 74th Congress, 1st Session, March 20, 1935, 4–5.

153. Ibid.
154. William Paul Robinson, *Groundwater Contamination in a Poor and Minority Community: The South Valley of Albuquerque, New Mexico*, Southwest Research and Information Center, 1985.
155. Price, *The Orphaned Land*.
156. Carolyn Carlson, "Neighbors Target Crime, Pollution," *Albuquerque Journal*, April 2, 2004.
157. Larry Rulison, "Living with Fear: A Chip Fab's Health Scare: Residents of a New Mexico Town Claim the Chemicals Used by Semiconductor Industry Are Making Them Ill," *Times Union* (Albany, NY), 2010, accessed March 9, 2017, https://www.timesunion.com/business/article/Living-with-fear-A-chip-fab-s-health-scare-709903.php.
158. South Valley Regional Association of Acequias, "Santolina Master Plan in the West Mesa," accessed March 9, 2017, http://www.southvalleyacequias.org.
159. Ibid.
160. Ibid.
161. Alasdair Baverstock, "A Bloody War for Water in Mexico," *Vice News*, 2014, March 9, 2017, https://news.vice.com/article/a-bloody-war-for-water-in-mexico.
162. Concerned Citizens for Nuclear Safety, "Kirtland Overstated Technical Conclusions about Jet Fuel Plume," 2016, accessed March 24, 2017, http://nuclearactive.org/kirtland-overstated-technical-conclusions-about-jet-fuel-plume.
163. Michael E. Campana, Olen Paul Matthews, Richard M. DeSimone and Doris J. DeSimone, *Policy Conflicts and Sustainable Water Resources Development in New Mexico's Rio Grande Basin*, Publication No. WRP-2, University of New Mexico, Water Resources Program, February 2000.
164. South Valley Regional Association of Acequias, "Santolina Master Plan."
165. Concerned Citizens for Nuclear Safety. 2016. *Kirtland Overstated Technical Conclusions about Jet Fuel Plume*.http://nuclearactive.org/kirtland-overstated-technical-conclusions-about-jet-fuel-plume/. Accessed: March 24, 2017.
166. Rulison, "Living with Fear."
167. Ibid.

Chapter 8

168. Verna Williamson, "Spirituality and the Native Earth," *Talking Leaves* 9, no. 2: A Sense of Place (Summer/Fall 1999).
169. Department of Interior, *Pueblo of Isleta Settlement and Natural Resources Restoration Act of 2006*, accessed March 9, 2018, https://www.congress.gov/109/plaws/publ379/PLAW-109publ379.pdf.
170. City-Data.com, U.S. Census Data, *Population Data for City of Albuquerque, New Mexico*, accessed March 9, 2017, http://www.city-data.com/city/Albuquerque-New-Mexico.html.
171. Campana, Matthews, DeSimone and DeSimone, *Policy Conflicts and Sustainable Water Resources Development*.
172. United States Court of Appeals for the Tenth Circuit, *City of Albuquerque & Carol Browner v. EPA & Isleta Pueblo*, 1996, accessed March 9, 2017, http://altlaw.org/v1/cases/1060742.
173. Ibid.
174. Ed Williams, *Small Tribe, Big River: Isleta Eyes Pollution in the Rio Grande*, 2015, accessed March 9, 2017, http://publichealthnm.org/2015/02/18/small-tribe-big-river-isleta-eyes-pollution-in-the-rio-grande.
175. Christopher Abeita, "Albuquerque Manager Admits Plant Was Not Prepared for Sewage Spill," Isleta Pueblo Politics, 2015, accessed March 9, 2017, http://isletapueblopolitics.com/tag/albuquerque-bernalillo-county-water-utility-authority.
176. Ibid.

Chapter 9

177. J.P. Bradbury, "Limnology of Zuni Salt Lake, New Mexico," *Geological Society of America Bulletin* 82, no. 2 (1971): 379-398.
178. Winona LaDuke, "The Salt Woman and the Coal Mine," *Sierra Magazine* (November 2002), accessed March 9, 2017, http://sacredland.org/zuni-salt-lake-united-states.
179. A:shiwi A:wan Museum & Heritage Center, "Chronological History of Zuni," accessed March 9, 2017.
180. E. Richard Hart, ed., *Zuni and the Courts: A Struggle for Sovereign Land Rights* (Lawrence: University of Kansas, 1995).
181. LaDuke, "The Salt Woman."
182. "Sacred Land Under Siege," *Santa Fe New Mexican*, January 7, 2001.

183. Sacred Land Film Project, *Zuni Salt Lake*. Report by Amy Corbin, Earth Island Institute, posted on November 1, 2003.
184. Robert Struckman, "Salt Woman Confronts a Coal Mine: Zuni Defend Its Sacred Salt Lake from a Proposed Strip Mine," *High Country News*, October 8, 2001.
185. Marley Shebala, "Speakers: Zuni Salt Lake Is Not for Sale," National Association of Tribal Historic Preservation Officers, accessed March 24, 2017, http://www.nathpo.org/News/Sacred_Sites/News-Sacred_Sites62.htm.
186. William A. Dodge, *Black Rock: A Zuni Cultural Landscape and the Meaning of Place* (Jackson: University of Mississippi Press, 2007).

Chapter 10

187. City of Santa Fe Water Culture (organization), www.waterculture.org.
188. City of Santa Fe, accessed May 30, 2017, https://www.santafenm.gov.
189. Susan Hazen-Hammond, *A Short History of Santa Fe* (San Francisco, CA: Lexikos, 1988).
190. Ibid.
191. Oral story shared by Governor Gil Vigil, Tesuque Pueblo, September 9, 2016.
192. The Cathedral Basilica of St. Francis of Assisi, "Our Parish History," Santa Fe, New Mexico, accessed May 30, 2017, https://www.cbsfa.org/parish-life/about.
193. City of Santa Fe, "Watershed History," https://www.santafenm.gov/upper_watershed_history Accessed: June 6, 2017.
194. Ibid.
195. Santa Fe River Group Project, "Santa Fe River," accessed May 30, 2017, https://riverrestoration.wikispaces.com/Santa+Fe+River+Group+Project.
196. City of Santa Fe, *Watershed History*.
197. City of Santa Fe, "City Begins 'Living River Flow' Releases into Santa Fe River," accessed June 6, 2017, http://www.santafenm.gov/news/detail/city_begins_living_river_flow_releases_into_santa_fe_river.
198. Brian Handwerk, "Santa Fe Tops 2007 List of Most Endangered Rivers," *National Geographic* (2007), accessed June 6, 2007, http://news.nationalgeographic.com/news/2007/04/070418-ten-rivers.html.
199. Middle Rio Grande Conservancy, "The Rio Grande: A Ribbon of Life and Tradition," accessed July 2, 2017, http://mrgcd.com/History.aspx.

200. Dictionary.com, "Gentrification," accessed June 20, 2017, http://www.dictionary.com/browse/gentrification.
201. Chris Wilson, *The Myth of Santa Fe: Creating a Modern Regional Tradition* (Albuquerque: University of New Mexico Press, 1997).
202. UrbanDictionary.com, "Fanta Se," accessed June 20, 2017, https://www.urbandictionary.com/define.php?term=Fanta%20Se.

Chapter 11

203. Middle Rio Grande Conservancy, "The Rio Grande."
204. Susan Montoya Bryan, "Feds Issue Warning as Group Fights for More Rio Grande Water," Associated Press, 2016, accessed January 12, 2000.
205. Neena Satija, "Despite Efforts, the Rio Grande Is One Dirty Border," *All Things Considered*, accessed October 22, 2013, http://www.npr.org/2013/10/22/239631549/despite-efforts-rio-grande-river-is-one-dirty-border.
206. Jim Earhart, "Binational Study Regarding the Presence of Toxic Substances in the Rio Grande/Rio Bravo and Its Tributaries Along the Boundary Between the United States and Mexico," EPA Region 6, Toxic Substances Study—Questions and Answers, 1994, accessed March 11, 2017, https://www.epa.gov/nscep.
207. Russell Gold, "Laredo Water Is Toxic," *San Antonio Express-News*, 2000, accessed January 12, 2000,
208. Office of the State Engineer, *Rio Grande Compact*, accessed March 11, 2017, http://www.ose.state.nm.us/Compacts/RioGrande/isc_RioGrande.php.
209. Cadence Mertz, "Dirty War Continues—Illegal Dumpsite Found at Riverside," *Laredo Morning Times*, 1999, accessed January 12, 2000.
210. Montoya Bryan, "Feds Issue Warning."
211. Laura Paskus, "Does the Fate of the Silvery Minnow Foretell the Rio Grande's Future? Biologists Go to Great Length to Keep the Fish Alive, But It's Nearly Extinct in the Wild," *High Country News*, 2015, accessed January 12, 2017, https://www.hcn.org/issues/47.13/does-the-fate-of-the-silvery-minnow-foretell-the-future-of-the-rio-grande.
212. Ibid.
213. Office of the State Engineer. *Rio Grande Compact*.
214. Natalie Ballew, Benhi Bolhassani, Amelia Koplos, John Montgomery and Michael O'Connor, "Sharing Water in the Rio Grande," *Texas Water*

Policy, 2015, accessed March 11, 2017, http://www.texaswaterpolicy.com/blog/2015/1/23/sharing-water-in-the-rio-grande.

215. Ibid.

216. Laura Paskus, "Next Stop for Texas-NM Water Dispute: Supreme Court," *New Mexico Political Report*, 2017, accessed February 10, 2017, http://nmpoliticalreport.com/151659/next-stop-for-texas-nm-water-dispute-supreme-court.

217. Ibid.

Chapter 12

218. U.S. Environmental Protection Agency, "Superfund Timeline," accessed July 2010, https://www.epa.gov/superfund/superfund-history.

219. "Map of Superfund Sites (New Mexico),: *Dreaming New Mexico*, accessed February 10, 2017, http://www.dreamingnewmexico.org/energy/environmental-justice.

220. City Data, "Farmington, New Mexico," accessed March 6, 2017, http://www.city-data.com/city/Farmington-New-Mexico.html.

Chapter 13

221. City of Santa Fe Waste Management, "Illegal Dumping," accessed March 6, 2017, http://www.sfswma.org/trash/illegal-dumping.

222. Megan Geuss, "Landfill Excavation Unearths Years of Crushed Atari Treasure," ARS-technica, Gaming & Culture, 2014, accessed March 20, 2017, http://arstechnica.com/gaming/2014/04/landfill-excavation-unearths-years-of-crushed-atari-treasure.

223. Eliana Aubin, "Letter to the Editor: Dumping an Unwanted Dog or Cat Is a Fourth Degree Felony in New Mexico," Sierra County Humane Society, August 7, 2011, accessed March 20, 2017, https://www.facebook.com/notes/desert-haven-animal-refuge/dumping-an-unwanted-dog-or-cat-is-a-fourth-degree-felony-in-new-mexico/10150292032073257.

224. Daniels, Thomas J. Public Health Reports, Vol. 101, No.1: 1986. *A Study of Dog Bites on the Navajo Nation.*

225. Jacqueline, Vaughn, "Rez Dog Problem Remains Unsolved," Tuba City Humane Society, 2016, accessed March 20, 2017, http://tubacityhumanesociety.org/wp-content/uploads/2016/08/Rez_Dog_article_in_Flagstaff-Sedona_Dog_magazine03-0416.pdf.

226. Ibid.
227. Hannah Grover, "Shiprock Couple, Navajo Nation Animal Control Team Up to Help Injured Dog," *Farmington Daily Times*, 2014, accessed March 20, 2017.
228. Lauren Villagran, "Las Cruces Bureau: Nearly 40 Dead Coyotes Dumped Near Las Cruces," *Albuquerque Journal*, 2015, accessed March 20, 2017, https://www.abqjournal.com/519815/dead-coyotes-dumped-near-las-cruces.html.
229. Ibid.
230. Cole Miller, "Dead Animals, Trash Causing Stink," KRQE News 13, 2014, accessed March 20, 2017, http://krqe.com/2014/02/26/dead-animals-trash-causing-stink.
231. Chris McKee, "State Failed to Enforce Massive Tire Dump Clean-up, Will Restart Effort," KRQE News 13, 2016, accessed March 20, 2017, http://krqe.com/2016/09/22/state-failed-to-enforce-massive-tire-dump-clean-up-will-restart-effort.
232. Ibid.
233. KRQE News 13, "Dumping on the Rise in ABQ Arroyos," 2014, accessed March 20, 2017, http://krqe.com/2014/08/27/dumping-on-the-rise-in-abq-arroyos/#jp-carousel-46493.
234. Margaret Wright, "Illegal Dumping Threatens NM Sensitive Sites and Waterways," Associated Press, 2015, accessed April 3, 2017, http://www.washingtontimes.com/news/2015/jul/8/illegal-dumping-threatens-nm-sensitive-sites-and-w.
235. Ibid.
236. Ibid.
237. Paskus, "Does the Fate of the Silvery Minnow Foretell the Rio Grande's Future?"
238. New Mexico Acequia Association, *Mayordomo Project*, 2017, accessed April 3, 2017, https://lasacequias.org.
239. Ibid.

Chapter 14

240. Philip L. Martin, "Immigration to the United States," *Farm and Migrant Workers*, 2015, accessed May 5, 2017, http://immigrationtounitedstates.org/491-farm-and-migrant-workers.html.
241. Ibid.

242. Avaraham Astor, Research Gate, "Unauthorized Immigration, Securitization, and the Making of Operation Wetback," Latino Studies 7 (2009): 5–29.
243. Martin, "Immigration to the United States."
244. Ibid.
245. Keith S. Delaplane, the Bracero Archive, accessed May 5, 2017, http://braceroarchive.org.
246. Reies López Tijerina, *They Called Me "King Tiger": My Struggle for the Land and Our Rights*, translated and edited by José Ángel Gutiérrez (Houston, TX: Arte Público Press, 2000).
247. *Food Chains* (film), initial release, February 10, 2014, accessed May 5, 2017, http://www.foodchainsfilm.com.
248. Martin, "Immigration to the United States."
249. Ibid.
250. Esther Yu His Lee, "Farm Workers Forced to Work Dangerous Jobs for Little Pay," Think Progress, 2013, accessed December 15, 2016, https://thinkprogress.org/new-mexico-farm-workers-forced-to-work-dangerous-jobs-for-little-pay-8c15287a7a73#.kcdc2mgpf.
251. Keith S. Delaplane. "Pesticide Usage in the United States: History, Benefits, Risks, and Trends," Integrated Pest Management, University of Georgia, Athens, Georgia, March 1996, http://ipm.ncsu.edu/safety/factsheets/pestuse.pdf.
252. Ibid.
253. Pesticide Action Network, "Farmworkers Represent the Backbone and Marrow of Our Agricultural Economy. Yet This Group Is One of the Least Protected from On-the-Job Harms—Including Exposure to Pesticides," 2016, accessed December 15, 2016, http://www.panna.org/frontline-communities/farmworkers.
254. Ibid.
255. Ibid.
256. Aviva Shen, "Demand Protections from Pesticide Poisoning," Think Progress, 2013, accessed December 15, 2016, https://thinkprogress.org/farm-workers-demand-protections-from-pesticide-poisoning-7835c2d0a9e4#.m88u60l74.
257. Ibid.
258. Ibid.
259. Eduardo Gonzalez Jr., "Migrant Farm Workers: Our Nation's Invisible Population," Extension, Cornell University, Cooperative Extension, 2015, accessed December 15, 2016, http://articles.extension.org/pages/9960/migrant-farm-workers:-our-nations-invisible-population.

Chapter 15

260. Robert Torrez, "Ancient Peoples of New Mexico," New Mexico Office of the State Historian, accessed January 4, 2017, newmexicohistory.org/people/ancient-peoples-of-new-mexico.
261. Western Environmental Law, "BLM Defers Fracking Near New Mexico's Sacred Chaco Canyon" (press release), 2015, accessed January 4, 2017, http://www.westernlaw.org/article/blm-defers-fracking-near-new-mexico%E2%80%99s-sacred-chaco-canyon-press-release-1615.
262. Ibid.
263. Ibid.
264. Michael Coleman, "Zinke Cancels Chaco Canyon Lease Sale," *Albuquerque Journal*, 2018, https://www.abqjournal.com/1140105.

Conclusion

265. Viola Cordova, PhD, "When the Sacred Is Mundane" (lecture), Native American Philosophies, 2002, Oregon State University, Corvallis, Oregon.
266. Russell Thornton, "American Indian Holocaust and Survival: A Population History Since 1492," *American Indian Quarterly* 14, no. 1 (1990): 85–87, Lincoln: University of Nebraska Press.
267. Jessica R. Goodkind, Julia Meredith Hess, Beverly Gorman and Danielle P. Parker, "We're Still in a Struggle: Dine Resilience, Survival, Historical Trauma, and Healing," *Qualitative Health Research* 8 (2012): 1019–1036.
268. M. Brave Heart, "The Historical Trauma Response Among Natives and Its Relationship with Substance Abuse: A Lakota Illustration," *Journal of Psychoactive Drugs, Native American Postcolonial Psychology* 35, no. 1 (2003): 7–1, Albany: State University of New York.
269. Charles Anderson, "Heat and Violence," *Current Directions in Psychological Science* 10 (2001): 33–38.
270. Paul R. Epstein and Dan Ferber, *Changing Planet, Changing Health: How the Climate Crisis Threatens Our Health and What We Can Do About It* (Berkeley: University of California Press, 2011).
271. Susan Doherty and Thomas J. Clayton, "The Psychological Impacts of Global Climate Change," *American Psychologist* 66, no. 4 (2011): 265–276.
272. Chris Abbott, *An Uncertain Future: Law Enforcement, National Security and Climate Change* (London: Oxford Research Group, 2008).

Epilogue

273. City of Santa Fe, WildEarth Guardians and City of Santa Fe Reach Deal to Keep the Santa Fe River Flowing, https://santafenm.gov/news/wildearth-guardians-and-city-of-santa-fe-reach-deal-to-keep-the-santa-fe-river-flowing
274. Public News Service, NM Supporters of Gila River Protections Make 3rd Pitch in Nations Capital, https://www.publicnewsservice.org/2023-03-16/environment/nm-supporters-of-gila-river-protections-make-3rd-pitch-in-nations-capital/a83513-1.
275. KOAT Action News 7, Gila River Sees Record-Breaking Flood, https://www.koat.com/article/gila-river-new-mexico-flooding-2022/40969952.
276. AP News, US gets OK for cattle-shooting operation in New Mexico, https://apnews.com/article/business-climate-and-environment-forests-animals-8c8ad33483c8a4710ffe1a13c6e82661.
277. U.S. Environmental Protection Agency, Ten-Year Plan to Address Impacts of Uranium Contamination in the Navajo Nation, https://www.epa.gov/sites/default/files/2021-02/documents/nnaum-ten-year-plan-2021-01.pdf.
278. U.S. Environmental Protection Agency, Abandoned Mines Cleanup https://www.epa.gov/superfund/abandoned-mine-lands.
279. AP News, Army OKs $1.5M Settlement for Pollution at Former Depot, https://apnews.com/article/us-army-new-mexico-environment-environment-pollution-e798960be2bf00c756c8777da39b35b8.
280. Colorado Public Radio, New Mexico reaches $32M settlement over 2015 Gold King mine spill that polluted Colorado water, https://www.cpr.org/2022/06/17/new-mexico-settlement-over-2015-gold-king-mine-spill/.
281. CBS News, Coronavirus in Navajo Nation, https://www.cbsnews.com/video/coronavirus-in-navajo-nation/.
282. KRQE News, PNM brings drinking water to Navajo families, https://www.krqe.com/news/new-mexico/pnm-brings-sustainable-water-to-navajo-families/.
283. AP News, Oil Companies Join Fight Against US Nuclear Waste Facilities, https://apnews.com/general-news-8aeb26c6491afe060f6142f1c976fe1e.
284. New Mexico Museum of Natural History and Science Valles Caldera, Jemez Volcanic Field, https://www.nmnaturalhistory.org/volcanoes/valles-caldera-jemez-volcanic-field.

285. New Mexico Bureau of Geology & Mineral Resources, Our growing understanding of the the Rio Grande rift, https://geoinfo.nmt.edu/publications/periodicals/earthmatters/12/n2/em_v12_n2.pdf.
286. Earthquate Track, Earthquakes in Hobbs, New Mexico, United States, earthquaketrack.com.
287. U.S. Nuclear Regulatory Commission, Environmental Impact Statement For The Holtec International's License Application For A Consolidated Interim Storage Facility For Spent Nuclear Fuel In Lea County, New Mexico (NUREG-2237, Supplement 1), https://www.nrc.gov/reading-rm/doc-collections/nuregs/staff/sr2237/index.html.
288. Carlsbad Current-Argus, Western states join New Mexico in resisting nuclear waste storage, https://www.currentargus.com/story/news/2022/07/29/western-states-new-mexico-resisting-nuclear-waste-storage-texas-radiation-federal-fuel-energy-holtec/65382940007/.
289. Searchlight New Mexico, Bill to block Holtec nuclear waste site introduced in NM legislature, https://searchlightnm.org/nm-legislature-introduces-bill-to-block-holtec-nuclear-waste-site/.
290. Carlsbad Current-Argus, Western states join New Mexico in resisting nuclear waste storage, https://www.currentargus.com/story/news/2022/07/29/western-states-new-mexico-resisting-nuclear-waste-storage-texas-radiation-federal-fuel-energy-holtec/65382940007/.
291. U.S. Department of the Interior, Biden-Harris Administration Protects Chaco Region, Tribal Cultural Sites from Development, https://www.doi.gov/pressreleases/biden-harris-administration-protects-chaco-region-tribal-cultural-sites-development.

ADDITIONAL SOURCES

Beauvais, F. "Comparison of Drug Use Rates for Reservation Indian, Non-reservation Indian and Anglo Youth." *American Indian and Alaska Native Mental Health Research* 5 (1992): 14–31.

King, M., A. Smith and M. Gracey. "Indigenous Health Part 2: The Underlying Causes of the Health Gap." *Lancet* 374, no. 9683 (2009): 76–85.

Kirmayer, Lawrence J., Joseph P. Gone and Joshua Moses. 2014. *Rethinking Historical Trauma*. Impact Factor: 1.861, Ranking: Anthropology 17 out of 84, Psychiatry (SSCI) 63 out of 136.

National Institutes of Health. *Mental Health and Stress-Related Disorders*. Durham, NC: National Institute of Environment Health Services, 2015.

Nelson, Donald R., Colin Thor West and Timothy J. Finan. "Introduction to 'In Focus: Global Change and Adaptation in Local Places." *American Anthropologist* 111, no. 3 (2009): 271–274.

U.S. Department of Health and Human Services, 2004. "WHO Report. American Indian Adolescents." Accessed January 1–May 31, 2017. https://www.cdc.gov/violenceprevention/pdf/suicide-datasheet-a.pdf.

Additional Resources

The following resources are recommended if you are interested in learning more about Impacts on New Mexico communities and how you can help:

Amigos Bravos
http://mesaprietapetroglyphs.org
Beyond the Mesas
https://beyondthemesas.com/tag/navajo-water-rights
Bioneers
www.bioneers.org
http://www.dreamingnewmexico.org
Center of Excellence for Hazardous Materials Management
http://www.cehmm.org
Communities for Clean Water
www.ccwnewmexico.org
Concerned Citizens for Nuclear Safety
nuclearactive.org.
Conservation Voters of New Mexico
https://www.cvnm.org.
Diné C.A.R.E. (Diné Citizens Against Ruining Our Environment)
www.dine-care.org
Earth Care New Mexico
https://www.earthcarenm.org/about-us.
Eastern Navajo Diné Against Uranium Mining—MASE
https://swuraniumimpacts.org/eastern-navajo-dine-against-uranium-mining
Eight Northern Indian Pueblos Council
www.enipc.org
Environmental Education Association of New Mexico
https://eeanm.org/
The Environmental Justice Coalition of Northern New Mexico
https://issuu.com/EnvironmentalJusticeCNNM
Environment New Mexico Research and Policy Center
www.environmentnewmexicocenter.org
Frack Off Greater Chaco
www.frackoffchaco.org
Gila Conservation Coalition
www.gilaconservation.org
Honor the Earth—Navajo Homelands
http://www.honorearth.org/navajo_and_din_bii_kiya
Keep New Mexico Beautiful
www.knmb.org

MASE: Multicultural Alliance for a Safe Environment
http://swuraniumimpacts.org
Mesa Prieta Petroglyph Project
http://mesaprietapetroglyphs.org
National Environmental Justice Conference & Training
www.thenjc.org
New Energy Economy–Santa Fe
www.newenergyeconomy.org
New Mexico Acequia Commission
NMAC is a unique organization that honors our legacy of water governance while also working to adapt for the future. For nearly thirty years, NMAA has been responding to challenges through communication, training and education for acequias. Together, we have built a movement around the principle that "Water Is Life" (El Aqua es Vida) and that we are defenders of the precious waters that nurture our communities.
www.nmacequiacommission.state.nm.us
New Mexico Center on Law and Poverty
nmpovertylaw.org
New Mexico Environmental Health Organization
www.nmeha.org
New Mexico Environmental Law Center
www.nmelc.org
New Mexico Health Equity Partnership
nmhep.org
New Mexico Interfaith Power and Light
ttp://www/nm-ipl.org
New Mexico PIRG Fund
http://www.nmpirg.org
New Mexico Water Conservation Alliance
nmwca.org
New Mexico Wilderness Alliance
https://www.nmwild.org
No Nuclear Waste Aqui
onuclearwasteaqui.org
Nuclear Issues Study Group (NISG)
https://nuclearnewmexico.com
"On Poisioned Ground"—Chemical Heritage foundation
https://www.chemheritage.org/distillations/magazine/on-poisoned-ground

The Partnership for Earth Spirituality
www.earthspirituality.org
Project Coyote
www.projectcoyote.org
Sacred Poison (documentary)
https://www.youtube.com/watch?v=1wzbJYV1tu8&feature=youtu.be
Santa Fe Watershed Association
www.santafewatershed.org
Sierra Club–Rio Grande Chapter
www.riograndesierraclub.org
Southwest Organizing Project
https://www.swop.net
Southwest Research and Information Center
www.sric.org
TEWA Women United
tewawomenunited.org
Tularosa Basin Downwinders Consortium
https://www.trinitydownwinders.com
Uncovered: New Mexico's Farm Workers
https://www.youtube.com/watch?v=byxOOWB2-m8
University of New Mexico–Institute for the Study of Race and Social Justice Department
race.unm.edu
Water, Air & Land: A Sacred Trust
Documenting the Growing Assault on the Health of the Land of Enchantment
http://sacredtrustnm.org
WildEarth Guardians
www.wildearthguardians.org
Yellow Fever
http://www.yellowfeverfilm.com

ABOUT THE AUTHOR

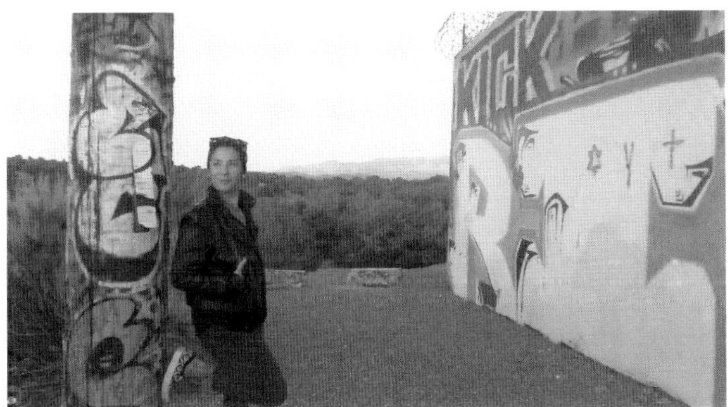

Credit: Frost Fowler.

Valerie Rangel has a masters of community regional planning degree that carried an emphasis in environmental and natural resource management, indigenous planning and public health, as well as an educational background in environmental science, Southwest history, Native American studies and cultural anthropology. She has taught college science courses and presently works as an environmental planning and public health assessment consultant and a community program manager for the state's community nonprofit foundation. She volunteers as a river steward and social justice activist. This book is a compilation of more than a decade of her research regarding social and environmental issues within the state; it is a history of land use in New Mexico highlighting case studies of communities in which environmental degradation has resulted in controversial battles concerning health disparities and sacred sites.